Steelmasters and Labor Reform, 1886–1923

Steelmasters and Labor Reform, 1886–1923

Gerald G. Eggert

UNIVERSITY OF PITTSBURGH PRESS

This volume was published with the cooperation of the Pennsylvania Historical and Museum Commission in its continuing attempt to preserve the history of the people of the Commonwealth.

Published by the University of Pittsburgh Press, Pittsburgh, Pa. 15260

Feffer and Simons, Inc., London
Manufactured in the United States of America

Library of Congress Cataloging in Publication Data

Eggert, Gerald G.
Steelmasters and labor reform, 1886–1923.

Bibliography: p. 199
Includes index.
1. Welfare work in industry—United States—History. 2. Steel industry and trade—United States—Personnel management—History. 3. Iron and steel workers—United States—History. 4. Industrial relations—United States—History. I. Title.
HD7269.I52U54 1981 331′.0472′0973 81-50636
ISBN 0-8229-3801-4 AACR2

To Jean:
Always patient, always loving, always there.

Contents

Acknowledgments

MANY PEOPLE have contributed in various ways to this study. Not the least helpful have been my colleagues and students at the Pennsylvania State University who have patiently listened to my thinking aloud as I worked on this project. I am especially grateful to Alice Hoffman of the Pennsylvania State University Labor Studies Department who first located the Dickson Papers and called them to my attention. Charles Mann, Director of the Rare Books Collection, and Ronald Filippelli, Director of the Historical Collections of the Pennsylvania State University Library, greatly facilitated my use of the Dickson and related collections. Two of my graduate students, Mark McCulloch and Barbara Carson Chunard, assisted me in my research, and an unpublished paper on the Johnstown Strike in 1919 by Robert M. Peles of the University of Pittsburgh was extremely helpful in directing me to fresh sources on that affair. The Central Fund for Research at the Pennsylvania State University provided financial aid for preparing the manuscript. Glen D. Kreider and William E. McCane of the Liberal Arts Data Laboratory generously assisted me in using the University's computer in preparing and editing the final version of the study. Professor Sidney Fine of the University of Michigan, Helmut Golatz, Head of the Department of Labor Studies at Penn State, and Irmgard Steinisch, at the time a doctoral candidate at the University of California, Berkeley, all have read the manuscript and offered helpful suggestions for improving it. Such errors of fact and interpretation as remain are my own.

Introduction

INCREASINGLY, American historians are recognizing that decisive changes in the internal structure of the United States took place between 1890 and World War I. Among other things, the period witnessed the appearance in modern form of many of the nation's most important private, public, and quasi-public institutions: corporations and business combines, trade and professional organizations, labor unions, governmental regulatory boards and commissions, and a host of other large-scale bureaucratic organizations.[1] Over the years a variety of labels have been affixed to various facets of this crucial transformation: "finance capitalism," "the managerial revolution," "scientific management," "welfare capitalism," "business unionism," "industrial democracy," "the regulatory state," "Progressivism," and more recently and more inclusively, "corporate liberalism." Each of these movements had roots in earlier eras, but all came to fruition in the quarter-century before World War I. And while some of the terms refer to specific financial, industrial, economic, social, or political developments, all were interlinked and broadly related.

The restructuring of the American iron and steel industry in that era illustrates the interrelationships of many of these movements. The wave of mergers in the industry between 1898 and 1909 offers examples of both vertical combination (bringing under single control all stages of an industrial process from the production of raw materials to the distribution of the finished product) and horizontal combination (unifying under single direction a number of once-independent firms engaged in the same line of business). The formation in 1901 of the United States Steel Corporation—a holding company—brought between 60 and 70 percent of the nation's steelmaking capacity under single direction and gave the industry its present oligopolistic structure. Although U.S. Steel produced only about 23 percent of the nation's raw steel by 1977, it and six other giant firms continued to dominate the industry.[2]

The role of the House of Morgan in creating, financing, and governing U.S. Steel provides a classic example of "finance capitalism" in operation. Bankers, financiers, and their lawyers reorganized the industry, succeeding to the power once held by steelmasters and imposing a new set of values on the industry. The key positions held by lawyers, bankers, professional managers, and other nonmanufacturers at U.S. Steel well illustrates the so-called managerial revolution. Founders of steel firms and self-taught steelmakers who had risen through the mills to positions of leadership were replaced by well-educated, highly trained specialists in management.

In the area of labor reform, U.S. Steel only partially followed the new trends. On the one hand, the corporation refused to extend tacit recognition to labor organizations, contrary to what some of the other employers did, and eliminated unions from its mills, operating on the open-shop (more accurately, the anti-union) principle. On the other hand, it imitated enlightened employers of the period by introducing a broad program of "welfare capitalism" (a series of humane, though paternalistic, benefits voluntarily provided to its employees). So far as means permitted, competing steel firms soon established similar programs of their own. Despite U.S. Steel's example, a few steel companies did experiment with "industrial democracy," at least in its weaker forms—company unions and employee representation plans. U.S. Steel itself carefully avoided all types of collective bargaining until the New Deal.

In recent years the term "corporate liberalism" has come into use as a label for much of the transformation under discussion. Originally employed by New Left revisionists as a reciprocating brush for tarring both big business and liberal politics, the term meant the conscious interaction of willing government officials and businessmen in the interest of establishing a corporate hegemony over the entire socio-politico-economic order.[3] The revisionists frequently overstated their case. Nevertheless, their examination of relationships and understandings among steel executives and federal administrations (to cite but one example) has helped overturn the simplistic traditional view that Progressivism consisted of liberal reformers using the state to correct the abuses of big business, thereby forcing corporations to serve the public interest.[4] More recent scholars have defined "corporate liberalism" in a broader, more constructive, and less pejorative sense. These writers no longer confine the term to the manipulation of the state in the interest of big business.

> They have seen the emergence of an "organizational sector," taking shape between the market and political [structures] and developing its own kind of discipline and rewards. In this sector men organized themselves by function, interest, or commodity, not by class or locality. They established new services and administrative networks to correct what were perceived as failures of the market system or of the political machinery; and seeking new sources of authority, they allowed managerial and technical elites to take over large areas of policy making and to erect and fortify their own bases of power. In this sector, moreover, private and public operations often fused into interlocked continuums. A new breed of private leaders sought to build state agencies that could render needed services without supplanting or threatening the new private institutions. And on the other side, a new breed of public officials looked to enlightened private organizations as the instrumentalities through which they could best advance the public interest.[5]

Although the general outlines of the transformation of American institutions are known, much research remains to be done on the details. How did the process take place in the different segments of society? What persons and forces shaped the process in each field of endeavor? What factors or persons delayed or speeded the process? What impact did the changes have on other groups?

This study undertakes to examine one aspect of the institutional transformation: labor reform in the steel industry during the late nineteenth and early twentieth centuries. Most of this development occurred in the private sector, government only infrequently becoming involved and then usually to press reforms on the industry. In particular, the study will explore the reform process among top steel officials. Where did the reforms originate? Who promoted and who resisted them? What were the motives and objectives of those involved, and what were the ultimate results? To focus so narrowly on one small part of the organizational transformation runs the risk of obscuring or even missing the "bigger picture." But it has the advantage of avoiding one of the chief faults of the broad approach, where change too often appears to be the product of certain irresistible, impersonal forces. Labor policy changes in the steel industry in the end were the product of a series of decisions made by leaders of the industry. Although these changes were obviously related to such forces as technology and economics, other leaders would not have reached identical solutions nor adopted the same policies, given the same authority and the same problems. In other words, there were alternatives to the courses pursued. We are better able to see how decisions came to be made, by whom, why, and with what consequences by considering those alternatives and why they were not adopted or why they failed.

For the period under discussion, labor reform in the steel industry was essentially reform from above. Unions, weak and waning in 1901 and all but eliminated from the industry by 1909, hovered in the background only as a remote threat to management. The conditions of labor in the industry cried out for reform. Many of the early reforms began at U.S. Steel, which is not surprising, given the company's position in the industry. The standards set by the corporation soon became the pattern for the industry at large.

The apparent assurance with which U.S. Steel introduced its various labor reforms over the years gave the impression that the program was carefully planned and willingly and generously bestowed on workmen as quickly as technology and prevailing economic conditions allowed. That impression was misleading. Some reforms did receive thoughtful study before being introduced. More often, however, they were hurried responses to threatened governmental investigations or muckraking exposes of conditions in the plants. Moreover, the leaders of U.S. Steel rarely acted with unanimity. Frequently they debated labor issues, split into factions, and sometimes restored to Machiavellian tactics to accomplish or thwart various programs. As often as not, debates over labor questions impinged upon other matters in dispute—U.S. Steel's commercial policy and, more important, control of the firm.

In its initial phase of labor reform, U.S. Steel voluntarily conferred certain benefits on workmen. While the trust did not invent welfare capitalism, its adoption of that type of program attracted wide attention.[6] In part this was because U.S. Steel advertised its every good deed to the fullest. But added significance was given to its policies by the mere size of the corporation and its dominant position in one of the nation's key industries. Included in U.S. Steel's program were an employees' stock purchase and bonus plan; improved company housing; home loans for employees; a safety drive; a scheme of liability compensation that provided payments to accident victims or their survivors; a system of retirement pensions; and health, sanitation, and comfort facilities in the mills and mill towns. Although some of the harder-headed managers of the corporation may have regarded these programs as unnecessary or even wasteful, they raised no strong objections. The reforms, after all, did not directly affect the area of their responsibility—steel production—and the expenses of the programs were not charged to their production cost schedules.

The second phase of labor reform in steel related to reducing hours of work. Periodically after 1900 the nation was shocked to learn from various investigations that despite shorter work hours in most other fields a large percentage of steelworkers continued to labor twelve-hour shifts and seven-day weeks. To reform-minded Americans such conditions were intolerable in the twentieth century. The battle for hours' reform involved a brief internal struggle among the chief officials of U.S. Steel, followed by a protracted public debate between the industry and its critics. Led by U.S. Steel, the industry dragged its feet; it repeatedly "studied" the problem, then delayed, resisted, and stalled as long as possible. Finally, with bad grace, spokesmen for the industry yielded to government pressures in 1923 and adopted the eight-hour shift.

The third aspect of labor-reform-from-the-top involved giving workmen a voice—albeit a carefully controlled voice—in the determination of conditions under which they would work. In steel the movement included neither encouragement nor recognition of regular labor unions. The principal union in the industry, the Amalgamated Association of Iron & Steel Workers, had once exercised considerable power because of the indispensible skills of puddlers, rollers, heaters, and others to the manufacture of iron. The evolving technology of steelmaking had reduced the demand for manual skills and had correspondingly weakened the position of skilled craftsmen in the industry. At Homestead in 1892, the Carnegie Steel Company broke the back of the Amalgamated, and strikes in 1901 and 1909 wiped out remaining pockets of unionism at U.S. Steel. Thereafter the corporation, and most of the industry as well, ran open shop until the New Deal.

A few steelmasters, however, came to believe that an institutionalized means of communication between employers and employees would serve the interests of both. Workmen could let off steam, air grievances, and make suggestions for improving work performances. Employers, in turn, could educate their men to the problems of management, defuse the just complaints of employees, and perhaps even generate a sense of common interest between workers and owners.

The push for worker representation did not take place at U.S. Steel but in two or three smaller steel firms during World War I. At one, the Midvale Steel & Ordnance Company, management invited workmen to participate in an employee representation plan (ERP) only because the federal government insisted that a labor dispute there, which threatened to disrupt vital war production, be resolved by some form of collective bargaining. The designer of the Midvale plan attempted to induce U.S. Steel and the rest of the industry to adopt similar plans but encountered only contempt and scorn. Not until collective bargaining became mandatory in 1933 did most steel companies find virtue in company unionism as a means for meeting their legal obligations while avoiding regular labor unions.

By chance one important steel executive, William Brown Dickson, played a significant part in all three stages of labor-reform-from-the-top in the steel industry between 1901 and 1923. Having begun as a common laborer in the mills at Homestead while still a boy, he experienced directly the working conditions that he later attempted to change. As assistant to Charles M. Schwab, president of U.S. Steel, then as second vice-president, and finally first vice-president of the corporation, Dickson helped to formulate and introduce a number of U.S. Steel's early employee benefits programs. In 1907 he moved center stage, introducing and leading the struggle against the seven-day work week and the twelve-hour shift. Confronting both superiors and fellow-managers, Dickson won a partial but temporary victory in 1910—the elimination of unnecessary Sunday labor. Having antagonized nearly all top officials of U.S. Steel in the process, he soon realized that his usefulness to the firm had ended, and he resigned in 1911. Even as he sought a new position, Dickson continued to snipe at U.S. Steel's hours' policy from outside the firm.

In 1915 Dickson became vice-president and treasurer of Midvale Steel & Ordnance Company, a newly formed combine created when the outbreak of World War I brought about an enormous market for American steel. By then he had a new concern: the lack of "democracy" in the mills. From his new station he pressed for adoption of a scheme of employee representation that would give workers a voice in working conditions. As at U.S. Steel, his superiors were unmoved. In 1918, however, the demand of the federal government that a labor dispute at Midvale be resolved by collective bargaining led to the introduction of an employee representation plan (ERP). Dickson's superiors yielded only because the plan avoided either recognizing or bargaining with a bona fide union. Lacking the full cooperation of both his superiors and the Midvale labor force, Dickson nursed the ERP for five years, trying to keep it alive and functioning. By 1922 Midvale's high wartime profits had turned to losses and several top officials fell ill or were considering retirement. Early the next year Midvale broke up, its properties passed to former competitors, and Dickson's "experiment in industrial democracy," as he called it, ended.

Between his second retirement at the age of fifty-eight in 1923 and his death in 1942, Dickson mulled over his experiences as a steel executive. "It would be difficult for anyone not intimately associated with my work," he admitted, "to find any striking evidence of my personality in the present business structure. . . . No process of steel making—no great labor-saving device, no notable discovery or invention, remain as a monument to my ability." His contribution, he concluded, had been the promotion of labor reform. Like Heine, the German poet whose work he admired, he had been but "a soldier in the war of liberation of humanity." [7]

Deciding to write his memoirs, Dickson assembled personal letters, minutes of important meetings, other corporate records, newsclippings, photographs, and memorandums. These he arranged in notebooks under appropriate chapter headings. Once he began a rough draft, the chapters on his boyhood, his early years in the mills, his courtship and marriage, and his rise to a junior partnership in Carnegie Steel came easily. When he reached the critical chapters on the struggle for labor reform, however, he faltered.[8] To relate that story seems to have raised internal conflicts that he could not resolve. On the one hand he remained loyal to the men with whom he rose to prominence in the industry; some were among his closest friends. Others—his "old boss" Andrew Carnegie and Judge Elbert H. Gary, the dominant figure at U.S. Steel during its first quarter-century of operation—he respected for their abilities, but they were the reactionary spirits who had blocked the reforms he saw as just and necessary, and his life's chief work. These chapters on the struggle for labor reform he never wrote.

What follows is not a biography of Dickson but an account of labor-reform-from-above in the steel industry between 1901 and 1923 and Dickson's important role in it. So far as possible the study attempts to reconstruct the internal struggle over reform as it took place in the boardrooms and executive offices of the U.S. Steel Corporation and the Midvale Steel & Ordnance Company. It not only deals with the complex and ambiguous motives of the reformers and their opponents but also makes clearer the scope, limitations, and ultimate shortcomings of welfare capitalism and employee representation as solutions to the labor problem in America.

Steelmasters and Labor Reform, 1886–1923

CHAPTER

1

Issues and Antagonists

THE CALL from the Pittsburgh headquarters of Carnegie Steel came in August 1889. The office there expected to be shorthanded for a week or so and wanted someone from the mill offices at Homestead, "preferably Billy Dickson," to help out temporarily. Charles M. Schwab, general superintendent of the Homestead plant, sent the telegram, but the special request for Dickson was the idea of one of Billy's friends, Charles L. Taylor.[1] The "temporary" transfer became permanent and marked the beginning of William Brown Dickson's rise to the highest managerial levels of the steel industry.

Until the call to Pittsburgh, which came when Dickson was twenty-three, his career had not been promising. He had engaged in manual labor at Homestead for five years and worked in the mill office as a clerk for three. Even after 1889 his progress, though steady, was not spectacular for another ten years. Then suddenly, at the turn of the century, position, increased authority, and wealth all flooded upon him. In January 1899, Andrew Carnegie named Dickson a junior partner in his firm. The next year, when Carnegie Steel became a corporation, Dickson was elected to the board of directors and appointed assistant to Schwab, who was by then the president of the company. Even as these events were taking place, a giant holding company—the United States Steel Corporation—came into being and absorbed Carnegie Steel. Schwab, chosen as president of the new steel trust, invited Dickson to continue as his assistant, and the next year Dickson had the title of second vice-president conferred upon him.[2]

Above and beyond his rapidly growing salary, Dickson acquired considerable wealth in the process. The junior partnership brought with it a one-ninth of one percent share in Carnegie Steel, carried on the books of the firm at $27,778.78. Its actual value probably exceeded ten times that amount. When the firm incorporated, Dickson converted his partnership share into 147 shares of Carnegie Company stock (with a par value of $1000 each) and $150,000 worth of company bonds. Then, in 1901, Dickson exchanged his Carnegie stock, with a par value of $147,000, for over $430,000 worth of common and preferred shares of U.S. Steel. He exchanged his $150,000 worth of Carnegie bonds for an equal value of U.S. Steel bonds.[3] By age thirty-six, Dickson was well on his way to being a millionaire.

At U.S. Steel, Dickson, a product of Carnegie methods and ideas, soon found himself caught up in a wide-ranging struggle for power. On one side were the men with whom he had risen to prominence at Carnegie Steel, men whose unmatched expertise in steelmaking won them positions of leadership in the new trust. Opposing them was a group of rising professional managers—lawyers and bankers by training—who represented the interests of J. P. Morgan, the nation's foremost investment banker and the chief financier behind U.S. Steel. For a decade the practical steelmen and lawyer-bankers were obliged to share power. During that period they vied with one another over the corporation's labor policies, commercial policy, the degree to which it should centralize control over the subsidiary companies, and, ultimately, who, under Morgan, would direct the destinies of the trust.

Dickson sided with the steelmen on most of the issues in dispute, the exception being labor policy. The lawyer-banker faction advocated conciliation of labor and the introduction of improved working conditions. Dickson's steelmaker colleagues opposed such reforms as unnecessary pampering of the men—and Dickson sharply disagreed. As a boy in the mills he had sworn that if ever given the chance he would attempt to reform working conditions. And so, without breaking with his usual allies, he worked closely with the lawyer-banker faction in planning and instituting U.S. Steel's early programs of welfare capitalism.

* * *

As with so much else in the history of steel, the conditions in the industry that made reform necessary in the twentieth century had become standard in the mills of Andrew Carnegie during the closing decades of the nineteenth century.[4] The technologies of steelmaking, the labor conditions, and the managerial practices that underlay the conflict over labor policies at U.S. Steel early in the new century were developed in the Carnegie mills. There many of the leading steelmakers of the next generation—men who would wield great influence at U.S. Steel, Midvale Steel & Ordnance, and Bethlehem Steel—began their careers. Even the first hints of labor reform in the industry can be traced to Carnegie. It was not that he invented or pioneered in any of these areas, for most of the technology and new ideas that he applied were borrowed from others. Rather, his influence stemmed from the fact that he owned and operated the largest and most successful steel firm in the country. Moreover, being more articulate than most steelmakers, he talked and wrote extensively about his accomplishments and views, making himself one of the world's best known businessmen.

At bottom, the secret of Carnegie's success lay in his remarkable ability to cut costs, reduce prices, and "scoop the market." [5] This meant running his mills to capacity while watching for every possible saving. He systematically measured costs at every stage of production to single out areas most in need of special attention. He spared no expense in installing cost-cutting machinery wherever practicable. He built or otherwise acquired the most modern plants in the industry. He bought up the blast furnaces and ovens that supplied him with pig iron and coke, and the iron ore and coal fields behind them. Railroad and oreboat companies he cajoled, threatened, purchased, or supplanted with new ones to keep down transportation costs. Once he began to pull ahead and benefit from economies of scale, Carnegie increased specialization in his mills, plowed back profits into plant improvement, maintained large capital reserves to facilitate new purchases, and soon surpassed every rival. In good years his prices bested theirs; in times of distress, when they cut production or closed down, his mills ran full, his output sold for whatever it would bring, and firms that fell by the way were bought up.

To no small degree Carnegie's achievements were due to the men he employed to manage his works. With a sure eye for talent, he hired or teamed up with some of the best men in the field: Alexander L. Holley, who constructed the Edgar Thomson Works at Braddock; Captain William Jones, a superb steelmaker and driver of men who managed the Edgar Thomson from its opening in 1875 until his death in 1889; and Henry Clay Frick, as shrewd and competent a man as Carnegie, but icier, harsher, and all but completely indifferent to the sensitivities of the men he drove. Carnegie began in time to produce his own managers, selecting young men of unusual drive, ambition, and talent from among his own employees. These he tested and promoted through the ranks. To the degree that they succeeded, Carnegie succeeded too.

With Carnegie setting the pace, the iron and steel industry in the eighties and nineties dramatically shifted from ironmaking to the manufacture of steel. Steel production, in turn, became increasingly mechanized and was conducted on an ever larger scale. The manufacture of rails illustrates these changes. In 1874 American producers turned out 650 thousand tons of rails, 20 percent of which were steel.

By 1880 two-thirds of the 1.3 million tons of rails produced were steel. A decade later, with production nearing 2 million tons, 99 percent were steel. Total steel production first surpassed iron output in the United States in 1892; within twenty years the ratio of steel to iron production would stand at 14 to 1. During the nineties the manufacturers of basic steel, all of whom had begun in the iron business, began to dismantle their ironworks and to discharge skilled ironworkers. By 1900 puddlers, for example, rarely were found on the payrolls of any of these companies.[6]

Blast furnaces erected in the 1870s had stood from sixty-five to seventy-five feet high and started out making fifty tons of pig iron a day. By 1881 the output of these same furnaces had been driven up threefold. New furnaces built in the eighties and nineties reached ninety feet in height and turned out between 200 and 350 tons of pig iron daily.[7] Meanwhile, improved Bessemer and then open hearth furnaces consumed the growing output of pig iron, converting it into steel. Indeed, larger and more sophisticated machines at every stage in steel manufacturing stepped up output, lowered costs, and reduced the need for manpower. Promoting the same results was the rise of integrated steel plants. Instead of each process being carried on independently in separate mills, related processes were brought together at a single site, saving both time and transportation costs.

The changeover to steel had a profound impact on working conditions in the mills. Iron production, unsusceptible to mechanization, moved at a relatively relaxed pace. "Some iron takes longer than other iron," a puddler explained to a congressional investigating committee in 1883. A "heat" might take from one-and-a-half to two hours, and traditionally puddlers made only five heats during a twelve-hour shift. That amounted to between eight and ten hours of actual work. Blast furnaces, which turned iron ore into pig iron, were the only operation that of necessity ran continuously. Most other steps in ironmaking were geared to the puddler's heats. This allowed, perhaps, an hour for eating and time for brief rest periods during the shift. The usual workweek was six days.[8]

Skilled workmen (puddlers, heaters, rollers, moulders, and the like) were indispensable to ironmaking. Moreover, they were tightly organized, especially in the Pittsburgh area, in the Amalgamated Association of Iron & Steel Workers. At its peak in 1891, that union embraced over twenty-four thousand members (perhaps two-thirds of all eligible workmen), negotiated uniform wage scales in the Pittsburgh district each year, and through elaborate work rules generally protected members from employer pressures to speed up.[9] Unfortunately for the workers, the shift to steel undermined the Amalgamated.

To steelmasters the antiquated work rules prescribed by the Amalgamated for ironworkers were neither necessary nor desirable in steelmaking, particularly when they resulted in pauses or work breaks. Every unnecessary stoppage meant a loss of heat that would cost time and money to replace. How much better for ore to pass through the blast furnaces emerging as pig iron, proceed directly to Bessemer or open hearth furnaces to be converted into steel, and from there on to rolling mills to be shaped into slabs, plate, sheets, rails, or wire without pause. The pace of work in the mills should be dictated not by the traditional and irrelevlant skills of workmen, but by the capabilities of the machines. While it was still true that only blast furnaces had to operate continuously, efficiency called for all subsequent processes to be continuous as well. First to go were the idle stretches between shifts; next to disappear were the long lunch periods as men learned to eat on the run; then the rest breaks, and finally, Sundays.[10]

The Amalgamated failed to adapt to the changes that were taking place, clinging, for example, to a restricted membership of skilled workers who largely ignored the swelling armies of unskilled steelworkers. Although the union showed some willingness to compromise on the matter of work rules, it declined to move as quickly or as far as the steelmasters demanded. The union's waning influence, particularly after its defeat in the Homestead Strike in 1892, doomed steelworkers to a pace of work determined by the machines and their employers.

Over the years three abuses in particular drew the heaviest fire of critics of the industry: the twelve-hour shift, the seven-day workweek, and the long (or twenty-four hour) turn. A workday, of course, did not have to be divided into two shifts of twelve hours. Three eight-hour shifts—or four six-hour shifts for that matter—would have served the same end. Nonetheless, the twelve-hour shift, already standard in blast furnace operations, quickly spread to the converting and rolling mills. In part the twelve-hour workday was traditional, coming over from iron puddling, as already noted, and from the agrarian past when working from sunup to sundown was the rule. But by 1890, the overwhelming majority of industrial workmen had already moved to the ten-hour workday.[11] Unfortunately for steelworkers, continuous operations did not easily accommodate to ten-hour shifts.

The spread of the seven-day workweek from the blast furnaces to the other departments also grew out of cost and efficiency considerations; cooling of the steel or equipment only to be reheated increased the cost of production. As long as there were orders to fill, the cheapest and most efficient operation called for a continuous, twenty-four hour schedule with no breaks whatever from day to day, week to week, month to month. If orders slowed, the mills did not cut back on hours or days of operation but would shut down partially or totally until orders once more justified operating on the old twenty-four hour schedule.

The introduction of Sunday labor stirred opposition. Churchgoers in steel towns and some clergymen raised strong objections. Captain William Jones, Carnegie's celebrated steelmaster, knew how to deal with such outside meddling. He notified those "bigoted and sanctimonious cusses" that if they interfered he would retaliate "by promptly discharging any workman who belongs to their Churches and thereby get rid of the poorest and most worthless portion of our employees." [12] The threat of losing their jobs for refusing to work on Sundays kept most men in line. Some, however, needed little persuading. The increasing number of Eastern European immigrants in the mills tended to favor Sunday labor. The goal of most of these men was steady work to earn a stake of money, enough to return to their homes in Europe and buy land, enough to bring their families to the New World, or, for some, enough to buy land in this country. For such men any breaks, even Sundays, only delayed realization of their dreams. Living miserably and apart from their families, some had little more than claim to a bed for the twelve hours a day they were not working. Sunday labor was preferred to an idle day with nothing to do except, perhaps, spend precious money for drink or entertainment.[13] Even American-born workmen, once caught up in the seven-day workweek, found a one-seventh reduction in weekly income too steep a price to pay for a day of rest. When the six-day workweek was introduced (with the loss of one day's wages each week) at a later date, workers resisted and in some instances quit their jobs.[14]

The barbaric long turn, or twenty-four hour shift, was the means by which workmen rotated from night shift to day shift and back. A day shift worker, for example, would labor from six o'clock Sunday morning around the clock to six o'clock Monday morning. Then, after twelve hours off, he would report for work at six on Monday evening and every evening thereafter for the next two weeks. At the end of the two weeks, he would quit at six on Sunday morning and enjoy a twenty-four hour break before returning to work at six o'clock Monday morning to begin two weeks on the day shift. The brief holiday was all too short and the workmen too tired to enjoy it. "I get home about half past seven Sunday mornin' and go to bed as soon as I've had breakfast," one worker reported in 1908. "I get up about noon so as to get a bit of Sunday to enjoy, but I'm tired and sleepy all the afternoon." He had followed that schedule for twenty years. "Home is just a place where I eat and sleep," many workmen testified bitterly. "I live in the mills." [15] Although Carnegie never acknowledged it, clearly much of his success came at a frightful cost to his employees.

But success—beating out the competition and realizing high profits—never of itself completely satisfied Carnegie. As a recent biographer has observed, he also craved recognition "before the world as America's most enlightened and progressive employer of a mass labor force." [16] The obvious inconsistency of high profits and generosity to one's employees came nearest to resolution during the few years that Captain Jones served as Carnegie's chief lieutenant and favorably tempered his employer's labor policies. Competitiveness characterized Jones, serving both as his personal motivation and the means by which he extracted ever more from the men who worked under him. Carnegie regularly wagered with Jones that such-and-such a record could not be broken. Just as regularly, Jones took the bait, surpassed the record, and collected his reward from a delighted Carnegie.[17]

In a paper presented to the British Iron & Steel Institute in London in May 1881, Jones explained the role that competition played in setting production records. The "*esprit de corps* of the workmen after they get fairly warmed to the work," he noted, was most important. "As long as the record made by the works stands as the first, so long are they content to labor at a moderate rate, but let it be known that some rival establishment has beaten that record, and then there is no content until the rival's record is eclipsed."

Within individual work crews Jones found ethnic tensions useful. To his English hosts he mentioned crews made up of "representatives from England, Ireland, Scotland, Wales and all parts of Germany, Swedes, Hungarians, and a few French and Italians, with a small percentage of coloured workmen." [18] As he explained more candidly elsewhere, these groups, excluding Englishmen who were too imbued with unionism, and "judiciously mixed" with " 'Buckwheats'—young American country boys," produced "the most effective and tractable force you can find," far better than gangs made up of "a preponderance of one nationality." [19]

Improved technology was another major factor to which Jones attributed his production records. Jones thanked the British institute for its generosity in publishing "free to the whole world, valuable information embodying years of careful labor and thought." He and other American steel producers, he confessed, made good use of these publications: they "simply swallowed the information" and "selfishly devoted ourselves to beating you in output." But new methods of production, Jones observed, again raised the question of labor. He noted a reluctance on the part of workmen, even the most intelligent, to accept innovation. "I can here say," he declared, "that generally all the improvements introduced at [our] works have been condemned and opposed by the workmen." [20] For that reason, if no other, Jones and Carnegie's other managers opposed unions among their employees. The resistance of unorganized workmen to "necessary" change was difficult; to have to negotiate each change with a labor organization would be intolerable.

Jones, however, was not simply a severe taskmaster and driver of men. He also recognized the importance to production of men who were treated fairly. Carnegie, gnawed by fear that his competitors might be paying less for labor, repeatedly questioned Jones on the matter of wages. His lieutenant's answer was always the same: "Low wages does not always imply cheap labor. Good wages and good workmen I know to be cheap labor." [21] He insisted that reduced wages would bring no new production records. When, early in the 1880s, Jones persuaded Carnegie to experiment with eight-hour shifts in the rail mills at the Edgar Thomson Works, efficiency was his motive, not humane treatment of the men. Jones intended the men to move as fast and as continuously as the machines. Discovering that "it was entirely out of the question to expect flesh and blood to labor incessantly for twelve hours," he instead drove three crews to the limit for eight hours.[22] Since wages hovered only slightly above subsistence, the men and their dependent families could hardly have survived if their wages had been cut a third to match the reduction in hours of work. Carnegie split the difference, paying them, as he put it, 16⅔ percent "more" for eight hours of work. As he admitted, without citing amounts, the actual cost to the firm was considerably less because three crews of less weary men worked more efficiently than two.[23]

So long as Jones lived (he died in an accident in the mills in 1889), Carnegie went along with the formula of good wages for good work. Meanwhile, he developed a taste for writing and basked in the praise generated by his generally optimistic and liberal point of view. In an especially ebullient mood in 1886 (his firm was to realize a 60 percent profit on its capitalization that year), Carnegie published two articles on the labor question in *Forum* magazine.[24] Although Jones's influence is evident, Carnegie went much farther, espousing what for the time were extremely advanced positions on wages, hours, labor organizations, and strikes.

As to wages, Carnegie cited as a major cause of labor discontent the inequitable division of the fruits of production between capital and labor. Proposals for profit sharing, he observed, would not work because "only five in every hundred" who went into business succeeded. And many of those were in "an anxious and increasing struggle to keep their heads above water." Such firms had few profits to share. Enterprises cooperatively owned by the workmen held out even less promise of success. Workmen who wished to share profits, Carnegie suggested, should buy shares of stock; that would give them a share of profits in the form of dividends as well as a voice in the management of the firm.[25]

The prevailing method of determining wages in steel was too inflexible, Carnegie declared; by this method employers and employees agreed on a fixed tonnage rate that ran for several years. If steel prices soared during the term of the contract, workmen felt cheated. If prices fell, profits disappeared. Carnegie proposed substituting a "sliding scale." Under this system a low minimum rate, beneath which wages would not be permitted to sink, would be set. Actual wages, however, would presumably be considerably higher and would rise or fall as the selling price of steel fluctuated. Under this system both good and bad times would be shared equally by employer and employees.[26] The plan may have been somewhat less liberal in intent than Carnegie's description indicated. Since his commercial policy was to drive steel prices lower than those of his competitors, to the extent that he succeeded the sliding scale would assure him of simultaneous and uncontested reductions in his labor costs.[27]

On the subject of hours, Carnegie chided the then rampant eight-hour movement as premature. Iron and steel workers, and the employees of gas works, flour and paper mills, and breweries, all worked twelve or more hours a day, and sometimes seven days a week. Surely something needed to be done for these workers before agitating for enactment of the legal eight-hour day. Even so, Carnegie conceded that eventually, factories with continuous operations "should be operated with three sets of men, each working eight hours." Steel rail mills in this country, he pointed out, had successfully moved to this schedule. "I trust the time has gone by," he added, "when corporations can hope to work men fifteen or sixteen hours a day. And the time approaches, I hope, when it will be impossible in this country, to work men twelve hours a day continuously." [28]

On the matter of labor organizations, Carnegie declared that the right of working men to combine and form trade unions was "no less sacred" than the right of manufacturers to form associations. The point "must sooner or later be conceded." It was his experience, moreover, that unions benefited both capital and labor. They tended, for example, to bring forward as leaders "the ablest and best workmen." They also gave workmen a sounder understanding of capital-labor relations. In the long run this would mean that employers would have to pay "more deference" to workmen personally, to their rights, and "even to their opinions and prejudices." He admitted that more intelligent workers would insist on "a greater share of profits . . . in the day of prosperity," but they also would more readily understand and accept wage reductions in depressed periods.[29]

Carnegie, in effect, also called for collective bargaining. The "natural leaders" of the workmen should "confer freely" with employers to resolve common problems. "Peaceful arbitration" should be resorted to when negotiations failed, and awards should be made retroactive to the date of referral. A strike or lockout ("a ridiculous affair,"), therefore, ought never to occur, he observed.[30]

Should a strike break out, however, Carnegie proposed that employers react with discretion; singling out labor leaders for particular punishment made matters worse because workmen saw such actions as a threat to them all. Although he condemned violence in labor disputes, he conceded that "To expect that one dependent upon his daily wage for the necessaries of life will stand by peaceably and see a new man employed in his stead is to expect much." Far better for the employer to let the workers "remain idle and await the result of a dispute." The hiring of strike-breakers would secure "neither the best men as men, nor the best men as workers." Carnegie even cautioned against being "unduly alarmed" at frequent disputes between capital and labor. "Kept within legal limits," such disputes were "encouraging symptoms" of "the desire of the working man to better his condition." On that desire hung "all hopes of advancement of the masses."[31] Carnegie never put these liberal sentiments into practice in his mills and pursued exactly opposite policies within a few years. Some of the principles would be revived by Dickson and like-minded reformers in the first decades of the next century, but most would not be adopted until the New Deal, a half-century later.

With Jones's death in 1889 and the emergence of Henry Clay Frick as second-in-command in Carnegie's empire, profit making regained dominance over concern for workers' rights. Toward labor Frick had a "simple, uncomplicated attitude: to regard men as a commodity like any thing else used in manufacturing—something to be bartered for as cheaply as possible, to be used to its utmost capacity, and to be replaced by as inexpensive a substitute as was available."[32] Even before

Jones's death, Carnegie abandoned the eight-hour shift at the Edgar Thomson works and introduced the sliding scale. These changes were delayed at the Homestead mill only because of the strength of the Amalgamated Iron & Steel Workers union and the weakness of his general manager there.[33]

The Homestead Strike of 1892 exposed to the world the gulf between Carnegie's professed views on labor and the reality in his mills. Having left Frick to manage negotiations with the Amalgamated, Carnegie withdrew to Scotland, knowing and approving of Frick's determination to get rid of the Amalgamated and to institute the sliding scale. Frick violated almost every principle set forth in Carnegie's writings: he issued an ultimatum to the union, he did not bargain in good faith, he did not submit the issues at dispute to arbitration, and he called in an army of Pinkerton agents to protect strikebreakers. After a pitched battle, the shedding of blood, and military occupation of the town of Homestead, Frick achieved his (and Carnegie's) goals: the union was broken, and all organized resistance to whatever steps management might take to make production more efficient came to an end.[34]

Improved machinery had made the return of the twelve-hour shift practicable once more, if not humane. Workmen needed to do less lifting and shoving than before, and output increased dramatically. As matters turned out, the more efficient machinery soon undercut the sliding scale. As output doubled and redoubled, the scale threatened to overly reward workers, from management's point of view. First Carnegie lowered the minimum rate, then, late in 1892, abolished it. In 1894 Carnegie abandoned the sliding scale altogether.[35] Thereafter wages no longer depended on worker productivity or the selling price of steel; only on what the labor market would bear.

In the year of the strike, Carnegie and his partners realized a profit of approximately $110,000 per day every day of the year. The workmen fared less well: before the strike, four of Homestead's 3800 employees earned an average daily wage of between $7 and $7.60; 109 earned from $4 to $7; and about 800 received between $2.50 and $4. These figures included most of the skilled workmen. Of the remaining three-quarters of the labor force, nearly 1200 received between $1.68 and $2.50; the remaining 1625, on average, took home $1.40 or less for between ten and twelve hours of backbreaking work. A study of family budgets in the industry at the time showed that a steelworker on average earned $578.52 annually, supplemented by $85.04 earned by other members of the family. Of this total of $663.56, rent consumed 15.3 percent, food required 45.1 percent, clothing 19.5 percent, books and newspapers 1.2 percent, alcohol and tobacco 6.6 percent, and other expenses 10.4 percent. This left $100.06, or 15 percent of total family income as "surplus." Without the earnings of wives and children, the surplus amounted to $15.02 or 2.5 percent of the earnings of the steelworker. The study concluded that if a surplus was to be built up "it must be at the expense of some of the children." The savings were "quite respectable" and could provide ample insurance against want "provided they could go on growing from year to year." Experience showed, regrettably, "that periods of strikes, shut-downs, illness or misfortune soon dissipate the little pile." [36]

After the strike (and during the ensuing depression), wages at Homestead and the other Carnegie mills were cut further. In 1893, Schwab, newly appointed general manager at Homestead, slashed tonnage rates for some skilled workers by as much as 50 percent and wage-rates generally by 15 percent. For 1895 he set himself the goal of sweating another $0.5 million "waste" out of manufacturing costs. Needless to say, much of the "waste" came out of wages. Even with the return of prosperity, a 40 percent increase in output, and a ten-fold increase in profits over 1892, wages rose only about 10 percent.[37]

Critics have debated the relative efficiency of the twelve- and eight-hour shifts. Some have tried to have it both ways: not only was the eight-hour shift more humane, but improved efficiency made it cheaper. But this argument implies that Carnegie and his managers either did not know what they were doing or were determined to oppress their workmen regardless of added costs. When he abandoned the eight-hour shift, Carnegie gave as the reason that none of his competitors had moved to the shorter shift. Obviously, he would not have changed back if three shifts had proved cheaper than two, no matter what the policies of his competitors. Apparently the increased efficiency of less exhausted men on the eight-hour shifts had not matched the added cost in wages. Inasmuch as production gains increasingly came from improved machinery, the contribution of labor, fresh or weary, mattered comparatively less.[38]

At least one of Carnegie's lieutenants believed the two twelve-hour shifts were both cheaper and more efficient. Testifying in opposition to eight-hour day legislation in 1903, Alva C. Dinkey, then general superintendent at Homestead, told a Senate committee that three eight-hour shifts were substantially less efficient than two of twelve hours. For the six-month period ending in June 1892, the rolling mills at Homestead produced 24,300 tons of steel with eight-hour shifts. For the same period one year later, after the strike and restoration of the twelve-hour shift, the same plant rolled 31,500 tons, or nearly 30 percent more steel, without new machines or larger crews. In explanation, Dinkey suggested that because the men worked as teams they could move no faster than the slowest man. "The workmen get to know each other by a sort of instinct," he said, "and they go forward to develop the speed necessary. They can not develop that speed necessary in the first hours of the day [shift]."

"Is a man like a loosened watch spring which moves slowly at first and faster at the end of the uncoil?" a Senator asked.

"I think it is true," Dinkey replied, "that the men move faster in their work in the latter part of the day [shift]." [39]

After 1892 the labor pattern at Carnegie Steel was set. In continuous operations men worked twelve-hour shifts, seven days a week. In other departments the ten-hour shift, six-day week prevailed. Unorganized, the workmen were subject to whatever changes in their jobs management regarded as likely to improve productivity. Wages were set by what the market would bear. There would be few changes even after U.S. Steel took over Carnegie's properties in 1901.

* * *

During much of this period of transition in steelmaking, William Brown Dickson worked in the mills at Homestead. The precise combination of inherited characteristics and environmental influences that shaped his character and personality, that made him a successful steel executive, and that embued him with what proved to be a fatal reformist tendency cannot be known. The scant information about his early years, however, provides some clues. Born in Pittsburgh on November 6, 1865, Billy (as his friends always called him) spent his entire childhood and youth in Swissvale, a quiet farm village on the right bank of the Monongahela River, ten miles south of the Iron City. A "sylvan hamlet" surrounded by "fertile and well-kept farms, wooded hills, bountiful orchards, and waving fields of grain," Swissvale was the home of several well-to-do Pittsburgh businessmen including Billy's father, John Dickson.[40] Grandfather Thomas Dickson, a Scotch-Irish immigrant, had founded a successful coal business in Pittsburgh in the decades before the Civil War. By the time John took over the firm, the Dickson & Stewart Coal Company owned five mines in the area.

John Dickson, his Scots-born wife, Mary Ann McConnell Dickson, and their nine children (eventually there would be eleven) moved from Pittsburgh to Swissvale in 1866. Dickson & Stewart had recently acquired mining rights to all coal deposits underlying the hill district of the village, and John Dickson wished to supervise operations in person. The firm constructed a row of company houses and employed between fifty and sixty men and boys—chiefly Irish Catholic immigrants—to mine the coal and run it in cars down to the tracks of the Pennsylvania Railroad at river's edge.

Not long after the Dickson family's move, disaster struck twice. First John's health gave way, turning him into a semi-invalid. Then came financial ruin when an officer of the Nation Trust Company of Pittsburgh embezzled funds and bankrupted the firm. Dickson, an important stockholder in the bank, was liable under Pennsylvania law for debts of the firm up to twice the amount of his investment. Dickson's older children left school to find work. Billy, only nine years old at the time, worked summers and during school vacations as a fan tender in his father's mine. Two years later he quit school to take a job arranged for by his parents, running errands for a storekeeper in Pittsburgh. His earnings were slim: $3.00 a week, less $3.50 per month for rail fare to and from the city. Billy changed jobs frequently before settling down for two years of work at the Pittsburgh telephone exchange.

Despite their impoverished state and the necessity of their children leaving school to find work, the Dicksons struggled to uphold middle-class standards for their family: an earnest, success-oriented approach to life, regular attendance at the local Presbyterian church and Sunday school, and a love for good books and music. At Grandfather McConnell's knee, Billy had long since learned of Robert the Bruce and other heroic Scotsmen and memorized the poems of Bobby Burns. The Presbyterian pastor and his well-educated wife further aided in maintaining a "cultural atmosphere" by lending Billy books from their library and encouraging him in rhetorical skills at meetings of the young people of the church. And so, without formal schooling beyond the age of eleven, Billy remained an avid reader and lover of good literature and enjoyed music, singing, and public speaking all his life.[41]

The opening of the Dickson mine at Swissvale marked the beginning of that village's transformation from agriculture to industry. The experience of this town was not unique. By the 'seventies steel had started its conquest of the lower Monongahela Valley. Carnegie and his partners, for example, began construction of the giant new Edgar Thomson Works at Braddock's Falls, three miles south of Swissvale, in 1872. The Edgar Thomson's Bessemer furnaces poured their first steel in August 1875. As the effects of the Panic of 1873 waned, the demand for steel rails swelled, and Carnegie, unable to turn out steel fast enough, suddenly cut off the supply of ingots to various fabricators who depended on him for their raw material. In response, several of his customers incorporated the Pittsburgh Bessemer Steel Company and built their own plant for production of the steel they needed. The new firm erected the most modern and efficient plant of its size in America at Amity Homestead on the left bank of the Monongahela, directly opposite Swissvale. There, only fifteen months after breaking ground, the Homestead Mill began pouring steel on March 19, 1881. By summer's end it was turning out rails in competition with Carnegie.[42]

The opening of Homestead created a great demand for labor in the area, and young Dickson's Swissvale neighbors, friends, and former schoolmates were soon crossing the river to find jobs in the mills. Although Billy enjoyed working for the telephone company, the promise of higher wages and the saving of train fare to Pittsburgh lured him to Homestead. The fifteen-year-old boy first entered the mills on the evening of April 28, 1881—a date he faithfully commemorated each year for the rest of his life. Awestruck and a little afraid, he surveyed the glowing scene before him. Life in Swissvale, working in his father's dank mine, even running the dingy streets and alleys of Pittsburgh had in no way prepared him for what he now faced. The scale of the mill reduced the workers to the size of ants as they tended the flaming furnaces. Acrid, sulphorous stenches, roaring, hissing sounds, withering blasts of heat, and screeching, thudding noises from the machines engulfed him. Above the din a paymaster asked who had put his name on the rolls. Billy did not know. Since the plant was shorthanded, the technicality was overlooked and he was put to work.

At first Dickson spelled a former schoolmate who sat in a pulpit controlling the hydraulic valves that operated a giant crane. In the weeks and months that followed, Billy himself became a pulpit boy, running cranes in the converting mills and later in the blooming mills. He later recalled operating a cold saw in the rail mills on the day that President Garfield was shot. For a while he bossed a gang of steel loaders in the yards.[43]

Early in his career young Dickson encountered and developed a strong distaste for labor unions. Neither he nor his brother Robert, who also worked at Homestead, were old enough to belong to the Amalgamated Association of Iron & Steel Workers when they first began work. Attempts by that organization to unionize Homestead led to a strike early in 1882. Both boys continued to work. One morning patrolling strikers caught Robert on his way to the mills and beat him. Recognizing his assailants, Robert preferred charges, and the men were tried and sentenced to prison. Tensions ran high and those who continued to work found it necessary to live around the clock in the mills. Sleeping on a cot provided by the company, Billy kept a large club at hand for protection against the strikers who threatened to storm the plant. When the strike ended, the returning unionists kept up a steady harassment of those who had remained on the job. Once, union men with whom he was working deliberately threw a rail against Billy's leg, cutting it deeply and leaving a permanent scar.[44]

Continuing troubles with the Amalgamated and a sharp drop in the price of rails induced the owners of Homestead to sell their plant to Carnegie in 1883. Although the change of owners was to have a profound effect on both the steel industry and on Dickson's career, the immediate impact on the boy's life was negligible. For five years he worked at Homestead as a manual laborer. Arising at five (sometimes seven days a week), Billy hurried through breakfast and crossed the Monongahela in a small rowboat. After twelve hours in the mills he recrossed the river, arriving home about seven in the evening. Years later he learned how anxiously his mother had awaited his step, fearing some mishap in the mills or on the river during floodtime or when the ice was running.[45] In 1886 Dickson was shifted to clerical work in the mills. For three years he kept records for the yardmaster, served as a payroll clerk, and finally became assistant to the chief clerk. Less exhausted at the end of the day, he enrolled in night classes at Duff's Business College in Pittsburgh and studied bookkeeping, accounting, and penmanship.

By the age of twenty, Dickson had not yet decided on a career. A lowly clerk in the mill office at Homestead, he was not certain that the steel industry held much of a future for him. Frequently called on to assist the plant physician in giving first aid to men injured or burned in shop accidents, Dickson weighed the possibility of going into medicine. He had no money, however, and except for a few business courses, no formal education beyond early grammar school. Moreover, by this time his future wife had to be taken into consideration.

Since the age of thirteen, Dickson had been in love with Mary Bruce ("Mamie") Dickson, the orphaned daughter of a once prominent Pittsburgh banker who lived in Swissvale. Although sharing the same family name, the two were not related. The girl's guardians—two spinster aunts and three bachelor uncles—preferred that the couple remain unrelated. Despite his uncertain prospects, a secret engagement followed by a year's trial separation, the determined opposition of Mamie's family, and an older, more prosperous suitor for her hand, Billy (or Will, as she called him) persisted and won. In February 1888, when both were twenty-two, they married and began raising a family.[46]

The matter of Dickson's career soon took care of itself with the call to the Pittsburgh office in August 1889. He never returned to work in the mills, though over the years his various responsibilities took him on business to all parts of Carnegie's Pittsburgh kingdom. Work at the firm's headquarters threw Dickson into contact with a group of driving young men who were destined to play a major role in the steel industry over the next four decades. These were some of Carnegie's "Young Geniuses"—men who started in the mills as boys, had shown ambition, ability, and inventive or leadership potential, and were rewarded with rapid promotion to key managerial positions. For the few most successful, Carnegie reserved the ultimate prize: a junior partnership, with a share of ownership in the firm and potential admission to the world of great wealth. The careers of three of the most prominent—Charles M. Schwab, William E. Corey, and Alva C. Dinkey—were to interweave closely with that of Dickson in the years ahead.

At least superficially, the careers of Schwab, Corey, Dinkey, and Dickson were classic examples of the American "self-made-man" success story. All began in the mills as boys, and, as Horatio Alger would have expected, all worked hard (who didn't work hard in Carnegie's mills?), all showed diligence and ambition, and eventually each was perceived by his superiors to be several cuts above the average workman. New responsibilities and promotions followed, and, in time, each came to the attention of Carnegie himself. Eventually all four reached the topmost echelons of management in the steel industry.

As the rising young managers neared the top, the race became wildly competitive. At the same time, close cooperation among the contestants was essential since only through teamwork could any achieve recognition. For the few who succeeded, the contest was exhilarating:

> While we were all leading the strenous life with the Carnegie Co. in the old partnership days [one later wrote to Dickson], we were usually worried and cumbered with many cares, & often our zeal led us to do & say things which to an outsider would have seemed rude, but no outsider knew or could know the feeling of loyalty & kinship which existed between us all. We were one family in the sense of each one striving to make himself the most important member, & ready to defend all the rest against any charge from the outside, whether fair or unfair. Those days were to us all, an education along lines which always bring success & whose value will remain with us for life.[47]

A closer examination of the early careers of Schwab, Corey, Dinkey, and Dickson, however, reveals that much more than hard work and ambition accounted for their ultimate success.[48] The traditional rags-to-riches saga must be revised for these men. As boys they did not stand toe-to-toe at the starting line of the race of life with the thousands of others their age who were beginning in the mills. None of the four were "buckwheats," for instance, none were immigrants, and despite the poverty of two, none came from laboring-class backgrounds. All four families enjoyed—or struggled to maintain—a middle-class status. The Dicksons began as the most wealthy of the four families, and although they suffered considerably after the father's misfortunes in 1874 they managed to preserve much of their value system. The Coreys seem to have been continuously prosperous, the father operating a successful coal business in Braddock. The Schwabs, although only modestly comfortable, were never poor. A weaver by trade, the father had accumulated enough in savings to buy a livery stable in Loretto, Pennsylvania, when health problems forced him to find different work in a healthful mountain environment. The Dinkey family had to scrabble hardest to maintain status. The father, a combination farmer and railroad employee, died in an accident when Alva was only nine years old. The widow promptly sold the farm, moved with her six children to a large house in Braddock, and took in boarders. In part she chose the community because the nearby steel mills offered job prospects for her sons.

The relative wealth of the Schwabs and Coreys made it possible for their sons to enjoy far more education than was usual for young men entering the steel mills in that era. Charlie Schwab graduated from high school in Loretto and enrolled at St. Francis Academy, a local institution administered by the Franciscan order. Although he did not graduate from college, Schwab took a number of liberal arts courses and received instruction in perspective drawing, single and double entry bookkeeping, surveying, and engineering. Corey, who early in life decided to find a career—not just a job—in steel, remained in the public schools to the age of sixteen. He then began work as a general helper in the chemistry laboratory at the Edgar Thomson Works. Nights he spent either taking commercial courses at Duff's Business College in Pittsburgh or in reading up on chemistry and metallurgy—areas in which he became expert.

More in accord with the self-made-man formula, Dinkey and Dickson had little formal schooling. Dinkey quit school at thirteen to take a job as messenger and water boy at the Edgar Thomson Works. There he developed an interest in telegraphy and electricity and became self-taught in those fields. Dickson, as already noted, quit school at eleven. Except for the few night courses that he took at Duff's Business College, he educated himself. Unlike the others, he read widely in general literature, history, philosophy, and religion rather than on subjects of immediate practical value.

To some extent "pull" helped the careers of Schwab, Dinkey, and Dickson. Schwab, at seventeen, clerked in a store in Braddock where Captain Jones of the Edgar Thomson Works bought his cigars. Jones took a liking to the good-humored, self-confident clerk and pointed out that keeping a store did not offer much of a future to a bright young man; why not try the steel business? Schwab did, beginning at the bottom. His first job was as a dollar-a-day rod carrier for a team of surveyors who were laying out a row of new furnaces at the Edgar Thomson. Several other low-skill, low-pay jobs followed. Obviously if Schwab had loafed, or been content to remain at the bottom, his story would have ended there. On the other hand, most of the young men starting at the Edgar Thomson did not begin with Captain Jones, in effect, betting on their success. And Schwab worked knowing full well that Jones's eye was on his every move. Combining hard work with as much bluff as needed to get new and better assignments, and with a willingness to study at night what he would need for the next day's work, Schwab passed every test. Within six months Jones put him in charge of installing the row of new furnaces and introduced him to Carnegie. Shortly after, Jones gave Schwab the job of regularly reporting to Carnegie, who soon developed a high regard for his steelmaster's protégé.

Dinkey had advanced himself to the position of telegrapher at the Edgar Thomson by 1885 when he was nineteen. That same year he took a job as a machinist at the Pittsburgh Locomotive Works. After three years there he began work with one of the first firms to introduce commercial electricity in Pittsburgh. In 1889 he became secretary to the general superintendent of the Homestead Works, shortly before Schwab became the general superintendent. During the years when Alva first worked at the Edgar Thomson, Schwab met and fell in love with Dinkey's older sister, Rana. For three and a half years, while courting her, Schwab roomed at the Dinkey's boardinghouse and of course came to know Alva. Although there is no doubt that Dinkey possessed a number of valuable skills in his own right, the fact that he was Schwab's brother-in-law certainly aided his career after 1889, when Schwab moved swiftly to the top at Carnegie Steel. As already noted, Dickson received a leg up on his career when his friend Taylor sent out to Homestead for clerical assistance and specifically asked for him. None of this is to denigrate the efforts put forth by the four young men or to deny their obvious talents. But neither should these other factors be ignored in any discussion of the success they achieved.

Once their careers were under way, three of the four—Schwab, Corey, and Dinkey—rose to the top as drivers of men and producers of steel. Schwab rose highest first. His formula for success included never quarreling with his superiors, never resisting their orders, and making himself indispensible by mastering the details of the business. For example, perceiving the growing importance of chemistry to steelmaking, he taught himself as much of that science as applied to the work he was doing. Schwab constantly sought out ways to increase output and reduce costs.[49] In 1887, at the age of twenty-five, he became general superintendent of the Homestead works. Upon Jones's death in 1889, Carnegie called Schwab back to the Edgar Thomson as general superintendent. Following the debacle of the Homestead Strike, Schwab drew the unhappy assignment of again supervising Homestead. His duty was to restore morale and to shape the defeated and sullen workers once more into an effective labor force. His success made him the logical candidate, at age thirty-five, for the presidency of Carnegie Steel when it fell vacant in 1897. Four years later, when U.S. Steel absorbed the Carnegie properties, Schwab, then thirty-nine, became president of the world's first billion-dollar corporation.

Close on Schwab's heels came Corey. He left the chemical laboratory for stints as weighmaster and order clerk in the mill offices at the Edgar Thomson. In 1887 he accompanied Schwab to Homestead as superintendent of the open hearth department. Two years later, at age twenty-one, his authority was extended to include the plate mills and, in 1893, the armor plate department. Corey succeeded Schwab in 1897 as general superintendent at Homestead and, in 1901 when he was thirty-five, as president of Carnegie Steel, then a subsidiary of U.S. Steel.

Dinkey, who apparently had no love for office routine, switched from being secretary to the general superintendent at Homestead to the electrical department. By 1893 he had risen to superintendency of the power plant. On the side, earnings started to roll in from his inventiveness. He bought up and made profitable the electric light plant of the town of Homestead and organized a telephone system for Homestead and neighboring communities. He later sold both enterprises at a profit. When Corey became general superintendent at Homestead, Dinkey became his assistant. And, as Corey followed Schwab, Dinkey followed Corey, becoming general superintendent of Homestead in 1901 and president of Carnegie Steel in 1903 at the age of thirty-seven.

Dickson's career moved more slowly. Unlike his colleagues, his work at no point directly involved production or the driving of men. Instead, Dickson rose through the clerical departments of the firm. The eleven years that he spent at the headquarters of Carnegie Steel proved to be the most strenuous of his life. Sensing the opportunities opening before him, he was determined to succeed. His first assignment was as a clerk in the entry section of the ordering and shipping department. There he entered orders from customers on proper forms and transmitted them to the appropriate plants to be filled. Subsequent correspondence between the mills and the customers passed through his department. Almost immediately Dickson discovered one of the company's more tangled problems. Men like Schwab and Corey, "whose chief interest was in making tonnage records rather than in making deliveries to customers in accordance with contracts," ran the mills.[50] Corey, for example, would leaf through order books, tearing out those orders on which his department could roll the largest quantity of steel possible. Paying little or no attention to promised delivery dates, he would pass the orders on to his subordinates, directing that they be run "come hell or high water." Large orders of steel plate, piled just as they were rolled, soon began to clog the yards. When a customer inquired about a delivery, chaos reigned. Often the order had been rolled but lay inaccessible at the bottom of a stack somewhere in the yards. To accommodate the customer, the order would be rerolled, sometimes more than once, with all duplicates simply accumulating.

Dickson watched as Corey won his junior partnership with ever greater tonnage records. He suspected, however, that Corey would have lost out had not the advent of the Spanish-American War disposed of the wasteful snarl. The government, in a panic over the weakness of the navy, "sent men to Homestead and bought practically all the duplicate plates at high prices, thus cleaning up the entire mess."

The strain of being responsible for deliveries, but having no real authority over the mill superintendents, soon drove Hampden Tener, Dickson's immediate superior, to a breakdown. Dickson took over. Fortunately Tener, just prior to leaving for a rest, had worked out a system for weekly meetings with mill superintendents at which rolling schedules were set "by a compromise between mill efficiency and contract obligations." Dickson continued these sessions, playing the role of spokesman for the customers. When Schwab took over as president of Carnegie Steel, he expanded Tener's plan by establishing the Operating Department. This department, composed of approximately twenty-five high company officials and mill superintendents, met over lunch each Saturday. With Schwab presiding, operating problems were discussed thoroughly and resolved for the coming week. "This elbow touch," Dickson believed, promoted "understanding and fellowship" among the men, making possible the resolution of the company's inevitable problems through friendly cooperation.

Because of his familiarity with mill operations, Dickson was designated secretary of the Saturday meetings. The detailed minutes that he prepared brought him directly to the attention of Andrew Carnegie. At this stage of his career, Carnegie ran the business from New York City (and often from Skibo Castle in Scotland). He kept a tight grip on the firm through correspondence which, at his insistence, was frequent and highly detailed. When the others left the Saturday meetings to join their families, or take in a ball game, or simply to relax, Dickson spent his time writing up his reports. If a report was not on Carnegie's desk each Monday morning, Dickson received a telegram demanding an explanation. The reports, Carnegie told Schwab (who relayed the message to Dickson), gave him "a better understanding of the business" than the more formal minutes of the board of directors.

Meanwhile, Dickson continued to advance. Leaving the order and shipping department, he became chief clerk and later assistant general agent under Tener. Then, between 1899 and 1901, came the junior partnership, membership on the board of directors, appointment as Schwab's assistant and, finally, the second vice-presidency of U.S. Steel.

The difference in Dickson's career pattern from those of his colleagues suggests an explanation for the different ways that they later viewed labor questions. While Dickson spent several years as a laborer in the mills and was well into his twenties before beginning his climb to the top on the clerical side of the business, the others began driving men, supervising production, and moving upward at much younger ages. Schwab, Corey and Dinkey spent little time at the bottom. Within six months of entering the Edgar Thomson, Schwab was directing the installation of equipment. Corey within two years had risen to weighmaster and within five years was superintendent of a department. Both Schwab and Corey, while still in their teens, were directing the work of large crews of workmen. Data on Dinkey's career is scant, but he had become a telegrapher by age nineteen and soon left Carnegie's employ, temporarily, as a skilled machinist. Upon his return be began supervising workmen.

By contrast, Dickson spent eight years in the mills, five as a manual laborer and three as a clerk. His transfer to the Pittsburgh office came when he was twenty-three. By that age both Schwab and Corey were already superintending large departments, and Dinkey was secretary to the general superintendent of Homestead. Dickson labored at or near the bottom long enough to know the frustrations and numbing quality of such work. Long enough, too, to experience the fear of being trapped there as were so many of the men around him. Although

Dickson escaped before hope or ambition deadened, his memory bore permanent scars. He found the long hours of grinding labor and Sundays without rest or religion destructive of human qualities in workers. The twelve-hour shift and seven-day workweek to him were "twin relics of barbarism" to be abolished as quickly as possible.

Those who rose by setting ever greater production records in the mills they supervised could ill afford to concern themselves greatly about the workers. For these men the road to the top involved not only intense competition with one another, but a kind of competition with the men they directed. Success consisted of achieving new production records *while reducing production costs*. The "successful" manager in one way or another had to get more out of his men for less. Logic might suggest that if the way to get managers to achieve more was to hold out great rewards to them, then the way to get more out of workers should be to offer them greater rewards also. That was not the method Carnegie used. Workmen were spurred to greater efforts by reducing wage rates so that they had to produce more in order to maintain, not increase, their incomes. The reasons for approaching workmen and managers differently are not hard to discern. Added pay for workmen would substantially increase costs and reduce profits. Besides, it was unnecessary. Workmen rarely were in short supply, thanks to the nation's open immigration policy and the steady flow of farmers' sons to the factories. Moreover, when workers produced more the credit often belonged to improved machinery provided by the company. On the other hand, good managers were both valuable and scarce. If not tied by self-interest to a firm, they could easily move on other jobs at equal or better pay.

It would be surprising if the resulting dual system of restrictive wages for workmen and generous benefits for managers had no psychological impact on the managers. To the extent that they were able and willing to drive their men, they were regarded as successful. They could not, therefore, be overly sensitive to the effects their driving might have on the men. Fortunately for the managers, the growing gulf in income between themselves and their employees, with its attendant difference in lifestyles, made the adjustment easier. The workers, most of whom did not speak English, lived miserably. It was easy to see them less as fellow beings than as objects to be manipulated.[51]

That Dickson never held a position that required him to drive men to produce more may help explain why he had sympathy and concern rather than contempt or indifference for whose who worked in the mills, and why he spent much of his working life trying to institute reforms in working conditions. When he looked back he realized the extent to which luck and chance, not just his singular abilities, contributed to his rise. Success was the rare exception, not the rule. There but for the grace of God . . . (in this instance the intervention of his friend Taylor), Dickson might have remained with the great majority who spent their lives in empty toil. Once they reached the heights, Schwab, Corey, and Dinkey looked back nostalgically. The long hours of hard work at low pay they saw as beneficial and character-building. Their successes they attributed to the working out of the natural order, which, after much sifting and winnowing, brought the fittest to the fore. On those left behind—the "failures" in life—they wasted little sympathy.[52]

For Dickson and the other young Carnegie partners who came to sudden wealth in 1901, having a great deal of money to spend was a new experience. Carnegie had kept a tight rein over his "boys" by giving them wealth in a form they could not spend. Under the ironclad partnership agreement, the shares given to junior partners could not be sold except to the firm at book value, a mere fraction of their real value. Moreover, despite the grumbling of his older partners, Carnegie insisted on plowing most of his firm's enormous earnings into plant expansion and modernization rather than into dividends.

When Carnegie sold out, all the shares in Carnegie Steel were converted into negotiable securities—U.S. Steel stocks and bonds—or cash. In both Pittsburgh and New York, former Carnegie partners began raising palatial residences. Lavish parties, fancy yachts and automobiles, tours of Europe and the world, wholesale amassing of art treasures, and other ostentatious displays revealed that some of the poor boys who had made good were determined to enjoy their fortunes.[53] By contrast with some of his colleagues, Dickson handled his riches with restraint. Like the others, he built for himself and his family a splendid residence in Montclair, New Jersey, within easy commuting distance of his office at 71 Broadway, the New York City headquarters of U.S. Steel. The three-storied home, appraised at over $100,000, with furnishings valued at an additional $200,000, included twenty-eight rooms and nine baths. The first floor provided space for the family (living room, dining room, sun parlor, kitchen, and butler's pantry) and for entertaining (a large hallway, reception room, drawing room, and billiards room). The second floor was given over to separate bedrooms and baths for Dickson, his wife, their four daughters, and son. An office for Dickson, an extra bedroom, storage space, and servants' quarters occupied the third floor. In 1907 Dickson purchased a farm at Littleton, New Hampshire, where he built Highland Croft—a fine summer home—and a number of other buildings including a lodge called House-in-the-Woods.[54] In 1905 Dickson bought his first automobile, a Stevens-Duryea. His children attended the best schools and traveled abroad. Mamie Dickson collected bronzes and paintings.

Even so, Dickson managed his new riches with dignity and generosity. The family lived graciously, but quietly. They entertained, but their parties were not extravaganzas. The women of the household had only modest furs and jewelry, considering Dickson's wealth.[55] Raised as a child to love and respect good literature, Dickson began to buy books at the age of nine. Now, however, he assembled a large library, which he used constantly. The range and quality of his reading—from the classics to the most recent writers—gave him greater polish and erudition than many of his college-educated contemporaries. Dickson was forever copying poems and ideas from the books he read, weaving them into his conversations, correspondence, and public addresses. Some served as topics for discussion when he met with close friends for luncheon at the Banker's Club.[56]

Dickson also enjoyed writing, trying his hand at essays, allegories, and poetry. Most of his essays were serious. His poetry, however, ranged from doggerel verse for the amusement of family and friends on their birthdays and other celebrations to serious pieces: an expression of sorrow at the death of a friend's son, a sonnet to the beauty of nature, a bit of philosophy. At least two of his poems were published, but not in literary magazines. *Survey* published "The Interrupted Dream," which dealt with the fate of an immigrant laborer condemned for a crime, and, during World War I, *Manufacturers Record* accepted a poetic damnation of "William the Accursed," entitled "The Kaiser's Vision." [57]

Music and drama, too, were an important part of the Dickson family's life. Not only did they regularly attend the opera, musical comedies, and plays, but most of them also sang and played one or more musical instruments. Dickson spent many evenings playing the cello, in trios with musical friends, or accompanied by one or more of his children on the piano or violin. He also wrote short plays for his children to act.[58]

Dickson was always generous with his money. In 1901 he made gifts of $5,000 worth of U.S. Steel bonds to each of his brothers, sisters, a widowed sister-in-law, and his wife's uncle. In 1907, except for the uncle who meanwhile had died, Dickson established trusts of $5,000 each for the same persons. In addition to these gifts, amounting to $75,000, he financed excursions and vacations for his less affluent relatives, including a trip to Europe in 1905 for eight nieces, his two older daughters, and a chaperone.[59] Dickson also gave much time and money to civic affairs in both Montclair and Littleton.

* * *

Dickson came into his own just as the Carnegie Company reached its most prosperous period. Thanks to increased sweating of labor after the Homestead Strike and to new labor-saving machinery, steel production in Carnegie's plants jumped threefold between 1892 and 1900—from 878,000 tons to 2,870,000 tons. "Ashamed to tell you profits these days," Carnegie wrote a friend in 1899: "Prodigious!" The firm, capitalized at only $25 million prior to its incorporation in 1899, increased its profits from $4 million in 1892 to $7 million in 1897. Gains during the next three years were even more spectacular: $11.5 million in 1898, $21 million in 1899, and nearly $40 million in 1900.[60]

As profits soared, a new threat to Carnegie's position took shape. Although not yet evident, the steel industry was about to undergo the most thoroughgoing reorganization in its history. The long domination by Carnegie and his once-revolutionary competitive techniques was about to give way to a new era marked by consolidation and noncompetitive "harmony." The movement got under way in 1898. It began with a wave of mergers among steel fabricating companies, many of whom bought their basic steel from Carnegie. Just as Carnegie had reduced his costs by getting control of the sources of his raw materials, his customers proposed to reduce their costs in a similar fashion. By including producers of basic steel in their new combines, they would produce their own raw material and no longer depend on Carnegie. With his markets thus threatened, Carnegie struck back. He began laying plans for the construction of modern new fabricating mills of his own that would compete with the older mills belonging to his former customers.

This danger of an outbreak of intense competition distressed Wall Street financiers, and especially J. P. Morgan. Having financed the mergers of the fabricators, they now saw their profits endangered by destructive competition from the master-competitor of the industry. As it turned out, Schwab, the genial president of Carnegie Steel, had a solution: merge Carnegie's empire, the new fabricating combines, and assorted other steel firms into a single steel trust. Acting as broker between Carnegie (who from time to time hinted at retiring and devoting himself to full-time philanthropy) and Morgan (who ardently longed to see Carnegie out of steelmaking), Schwab, in effect, fathered the U.S. Steel Corporation and became its first president in early 1901.

The new holding company brought between 60 and 70 percent of the American steel industry under its control. Included in the merger were three important producers of basic steel (the Carnegie, Federal, and National Steel Companies), five leading fabricators (American Bridge, American Sheet Steel, American Steel & Wire, American Tin Plate, and National Tube), and a host of smaller properties. Each became a subsidiary company, maintaining its own identity and officers. Of course, each ultimately was subject to direction by the officers of U.S. Steel.[61]

As the trust took form, President Schwab moved to the firm's headquarters in New York City. Dickson accompanied him as his assistant. Another Carnegie partner, James Gayley, served as vice-president. Gayley, Henry Clay Frick (properly designated a Carnegie man in spite of his quarrel and break with Carnegie), and Thomas Morrison (another Carnegie partner), all sat on U.S. Steel's board of directors.[62] Other Carnegie men continued to manage Carnegie Steel: Corey was president of the firm, Dinkey general superintendent at Homestead.

If Schwab was "father" of the new trust, Morgan—as one congressman aptly put it—was its "Godfather." "Between conception and birth" of U.S. Steel, Morgan had regularly been consulted; he was "present at the birth and christening," and "after the thing was put on its feet," his views were sought out "in every hour of peril or crisis." [63] Although both the vast properties and expert steelmakers of Carnegie Steel were indispensible to the trust, Morgan did not intend U.S. Steel simply to be Carnegie Steel writ large. So he named Schwab president of the firm but at the same time made sure that a succession of Morgan agents and partners dominated the board of directors and the board's important executive and finance committees.

This group, for the most part bankers and lawyers by profession, was led by Judge Elbert H. Gary. The son of "prosperous and influential" parents, Gary was born on a farm near Wheaton, Illinois, in 1846.[64] In later life he fondly remembered the long hours of hard work that he experienced on the family farm and concluded that it had helped to mold his character and to prepare him for the long hours required by his duties as a principal officer of the steel trust.[65] Gary received his education at the Illinois Institute, a Methodist college his father had helped found. After serving two months in the Union Army near the close of the Civil War (Gary was only fifteen when the war began), he taught school until deciding on a career.

Gary read law in the office of his uncle, a local attorney, then enrolled for a year of study at the Union College of Law in Chicago. Once in practice, he flirted with politics, being elected the first mayor of Wheaton, and, later, two terms as judge of the county court. Gradually business law became his specialty. Hired as counsel by a number of railroad and industrial firms in the Chicago area, Gary soon found himself on the boards of directors of several corporations. Doing the legal work for the merger of five small firms into the Consolidated Steel & Wire Company gave him his initial contact with the steel industry.[66]

J. P. Morgan first took notice of Gary at a meeting where Gary told the banker that an action he proposed to take was not legal. According to Gary, Morgan replied, "I don't know as I want a lawyer to tell me what I cannot do." Gary, as he would many times in the future, asked Morgan what it was that he wished to accomplish and then proposed a way by which it could be done within the law.[67] Impressed, when Gary helped to create the American Steel & Wire Company, Morgan provided the financial backing. Next he employed Gary to organize the Federal Steel Company and made him its president.[68] Then, as U.S. Steel began to take shape, Morgan hired Gary to assist with the legal work and had him elected to the firm's board of directors.

Gary represented the growing influence of lawyers in corporate affairs. He knew little about steelmaking; indeed, one critic suggested that the first blast furnace Gary ever saw was in the hereafter.[69] But he did have certain legal and managerial talents needed by U.S. Steel. Gary's role at the trust between 1901 and World War I was threefold: as Morgan's representative, he helped to institutionalize a noncompetitive strategy for the steel industry; he protected U.S. Steel from assaults by the federal government under the antitrust laws; and, gradually, he consolidated full control over the trust (and, indirectly, substantial control over the industry as a whole) in his own hands. Assisted in the early years by George W. Perkins (a self-made insurance magnate and Morgan partner), Gary fought to end the Carnegie heritage in the steel industry. Gary and Perkins shared Morgan's repulsion of much that Carnegie stood for: fierce competitiveness, constant expansion of operations, ruthless cutting of costs, and repeated slashings of wage-rates and prices. Not only did they see these tactics as cruel, wasteful, and destructive, they regarded them as outmoded and no longer in the best interest of the industry or of society. The emerging economic order, these men believed, should rest on stability. Output should be geared to demand. Prices, wages, and profits should represent fair treatment of customers, laborers, and investors alike and should be steady. To secure and maintain this stability, competition among steel producers had to be reduced, if not eliminated.[70]

All of this would be the task of the rising new school of professional managers—men like Gary and Perkins. Although lacking expertise in any particular field of manufacturing, they saw themselves as competent to manage almost any type of enterprise. In administering a great "public" corporation such as U.S. Steel (public because it was owned by thousands of stockholders), these managers, unlike their predecessors who often managed firms that they owned personally, had to answer to that "public." As Perkins put it, the business decisions they made had to be "from the broadest possible standpoint of what is fair and right between the public's capital, which they represent, and the public's labor which they employ." [71]

The threat of antitrust action came from the Theodore Roosevelt administration, which had come to power in 1901 shortly after U.S. Steel's birth. Since the 1890s the courts and legal community had seen the Sherman Act as applying only to combinations that operated in interstate commerce, narrowly defined to exclude combines in manufacturing. But after the turn of the century, more and more lawyers came to believe that the law applied to monopolies and near monopolies in manufacturing as well.[72] The growing public outcry against the trusts in general, and Roosevelt's proposal to prosecute "bad" trusts while tolerating "good" ones, caused concern at the headquarters of U.S. Steel. Judge Gary's initiatives to ward off antitrust actions against the corporation lie outside the scope of this study; his determination to portray U.S. Steel before the public as a benevolent employer and "good" trust by introducing various welfare programs for the company's employees, however, will be dealt with below.

As for taking control of U.S. Steel, Gary had to move carefully. Carnegie men were firmly entrenched and their knowledge of steelmaking too important to the firm to be dispensed with. Although Gary would clearly advance his position during the first two years of the corporation's existence, his goal of complete domination would not be realized until 1911.

In advancing his various goals, Gary worked by indirection. He brought to the affairs of U.S. Steel an apparent openness: corporation reports were detailed and made public, annual stockholders' meetings were opened to the press, decisions within the firm were reached by lengthy discussions among the top managers. The men with whom Gary worked would eventually learn that what he seemed to say often turned out to be at odds with his actions. To say that he lied would be to oversimplify. Technically, Gary rarely told an untruth. His choice of words and phrases, however, contained semantic loopholes that frequently misled the unwary.

As the steelmakers, lawyers, and bankers moved into their respective offices at the headquarters of U.S. Steel, the jockeying for position began. How to deal with a threatened strike served as the first issue of substance to divide the factions. Underlying the decisions on this and most other matters, however, was the issue of who would control the corporation. Morgan's power was paramount. Unlike Carnegie, however, he had no interest whatever in the day-to-day operations of his properties. These matters he left to his lieutenants at U.S. Steel without clearly designating who should lead. No one was authorized to speak for him, none had frequent access to him, and all knew that he disliked having managerial questions brought to him for resolution. As a result, the early meetings of the managers and directors in some ways resembled a poker match. Bluffing, if done well, could win tricks. Power was conserved for use at the proper moment. There were even "wild" cards: the foibles and personal indiscretions of the principals that could be used to score points at critical junctures in the game. As some of the players were eliminated, Dickson moved ahead.

CHAPTER

2

Promises to Keep

THE LABOR CONDITIONS inherited by U.S. Steel in 1901 presented the officers of the new corporation with both an immediate practical problem and a longer-ranged project in labor reform and public relations. Of immediate concern was the threatened strike by the Amalgamated Association of Iron & Steel Workers unless the corporation agreed to extend union recognition to all its properties. Of more far-reaching concern were the problems stemming from public exposes of steelworking conditions; steelworkers, on the average, worked longer for less wages and endured more wretched living and working conditions than almost any other large group of laborers in the country. To the chagrin of U.S. Steel, muckraking journalists repeatedly made these facts known to the public.

In 1901 the Amalgamated had lodges functioning in perhaps a third of the mills belonging to U.S. Steel.[1] The shift from iron to steel in the late nineteenth century had severely weakened the union, which continued to limit its membership to skilled workmen. The consolidation movement in steel at the end of the century further imperiled the Amalgamated. Combines that included both union and nonunion plants could shift much of their business to their nonunion mills and thereby weaken or even destroy unions in the idled plants. The Amalgamated's only hope was to strike all mills in a combine at the same time, thereby, possibly, securing recognition. With the appearance of U.S. Steel in 1901, the Amalgamated correctly sensed that time was running out. Once the corporation perfected its organization, attempts to unionize it would be next to impossible.[2]

The executive committee of U.S. Steel's board of directors began discussing the labor question on April 20, 1901. Almost immediately the steelmakers and the lawyer-bankers separated into opposing factions. The seven-man executive committee divided about evenly during the first half of 1901. The lawyer-banker faction was led by Gary, who could usually count on the support of Daniel G. Reid (a banker turned steelman), Charles Steele (a lawyer, banker, and Morgan partner), and Percival Roberts, Jr. (a well-educated, second-generation steelmaker who inherited the family business). The steelmaker faction was led by Schwab. He usually had the backing of E. C. Converse and William Edenborn, both, like himself, self-made steelmasters.[3] Steele, however, frequently absented himself from meetings, and Roberts was given to taking erratic stands. The others, Reid, Converse, and Edenborn, though usually consistent, would desert their allies from time to time and vote the other side. The main accomplishment of the meetings from May to July of that first year was to sharpen the antagonisms between the two groups.

From the beginning, the steelmakers advocated a harsh stance toward unions and workers. As one of them put it, "I have always had one rule. If a workman sticks up his head, hit it." Gary objected. So long as he was chairman of U.S. Steel, no workman's head would be hit.[4] The lawyer-bankers proposed a more conciliatory attitude. Although opposed to unions, they intended to "treat the men fairly as individuals and give them good, liberal wages." [5]

The steelmakers suspected, not without justification, that the lawyer-bankers might not hold firm in a showdown with the Amalgamated. Whenever the lawyer-bankers spoke against unions, they invariably tempered their remarks with other issues: fairness to workmen, the need for peace at this juncture in the corporation's history, and financial considerations, to name but three. At a meeting of the executive committee on June 17, a vote was taken as to whether the committee approved "of fighting the thing out if a general strike were ordered." One member abstained, one voted in the negative, and one (apparently Roberts) explained that he would vote against recognizing a union anywhere but deplored U.S. Steel's inconsistency in recognizing unions in some of its mills and not in others. As he explained on another occasion, he was prepared to accept the union everywhere, or, if it was "a bad thing, wipe it out" everywhere. He preferred getting rid of the union but conceded that the time was not right.[6]

In spite of differences, the factions were able to agree on a general policy at their meeting of April 20. The corporation should "temporize for the next six months or year" until it had "fully established" itself.[7] Although the committee would debate tactics for dealing with the union well into July, the goal of all remained constant: a strike should be avoided so long as the corporation did not have its own house in order, and unionism should not be allowed to spread into new mills.

The challenge of the Amalgamated proved to be less an issue than a vehicle for the power struggle between the factions for control of the corporation. None of the minutes indicates the least fear of the union or concern as to who would win if a strike were called. The real issue was who would manage the affairs of the corporation: the executive committee, chaired by Gary, or President Schwab, working through his assistants and the presidents of the subsidiary companies.

The struggle began with consideration of tactics for dealing with the Amalgamated. At a meeting on April 20, Edenborn, one of the steelmakers, proposed that U.S. Steel adopt the position that it was only a stockholder in its subsidiary companies and exercised no direct control over them.[8] This would have the tactical advantage of forcing the Amalgamated to deal with the subsidiaries one by one rather than with U.S. Steel as the representative of all. To the steelmakers it had the further advantage of protecting the subsidiary companies from the overcentralizing tendencies of the executive committee.

Although the lawyer-bankers were most anxious to bring the independent-minded subsidiary presidents to heel, they found Edenborn's proposal an attractive tactic. It could be used to refute the frequent charge that U.S. Steel was a monolithic monopoly that dictated every business detail of its subsidiaries. Of more immediate importance, a strike involving all of its properties would probably follow if U.S. Steel agreed to speak for the subsidiaries and then refused the demand to recognize the union in all plants. A strike could possibly be avoided and the union kept in check by refusing to speak for the subsidiaries; U.S. Steel would confine the Amalgamated's effective bargaining power to the minority of mills where the union already had strength.

At the committee meeting on May 1, three of the steelmaker faction made a bid to give the subsidiaries autonomy in dealing with labor. The corporation, they proposed, should make no changes whatever in the existing labor relations of any of the subsidiary companies. If any other group (presumably the Amalgamated) sought to alter those relationships, the move should be resisted "by the ordinary means." From long experience, the steelmakers continued, they had learned that when certain problems arose in the mills they had to be "quickly disposed of on the spot with a firm hand" to prevent them from getting out of control. On occasion there was not even time for consultation with higher officials. In such circumstances, the subsidiary officers should be authorized to act and their actions should subsequently be upheld by the committee. Important issues, as a matter of course, would always first be brought to the attention of the executive committee.

The lawyer-bankers did not agree. At the "first intimation" that trouble was brewing, the president of the corporation should be "promptly and fully informed" by subsidiary officials and the executive committee should consider the matter and "give its advice." Subsidiary officers should "give no decision" until they had been instructed by "headquarters." The president of the corporation should inquire about labor problems in the subsidiary companies and spell out in detail the proper procedures to be followed by the officers of those companies. To give subsidiary officers authority to act on their own could lead the corporation as a whole into difficulties. Meanwhile, the corporation should "aim to learn what is fair treatment to the laboring men" and avoid the "great mistake" of appearing to have "adopted a policy of antagonism" toward labor. Above all else, the committee must "not lose sight of the financial interests of the corporation" or do anything that might prejudice those interests. The meeting ended inconclusively and further discussion was put over for a later session. Discussions the next day were equally inconclusive except that it was decided for the time being not to refer the matter to the whole board of directors.[9]

Within a few days Schwab was presented an opportunity to act on his own when trouble on the corporation's boats on the Great Lakes threatened to halt the movement of ore. There had been no time for prior consultation with the committee before acting, he reported. The committee retroactively approved his course. On May 28 Schwab reported that the subsidiary presidents were in session and that he had recommended to them "pursuance of the policy outlined at this committee by the president at previous discussions." No actual vote had been taken on that policy, he noted, but he "knew of no dissenting opinion of the committee as a whole" and so had proceeded accordingly. He now proposed and received a formal vote endorsing his actions "heretofore" and recommending pursuance of the "same policy in the future." The precise nature of that policy was not spelled out in the minutes of the meeting.

At the executive committee meeting on June 17 Schwab pressed for a clear policy statement advising subsidiary presidents how to respond if the Amalgamated presented them with proposed changes in wage scales or demands for union recognition in nonunion mills. He proposed that they be authorized to recognize the union and deal with proposals on scale but only for plants that were already unionized. Schwab assured the committee that all existing nonunion mills could be maintained as they were and subsidiary presidents could handle the matter simply by refusing to deal with the Amalgamated regarding them.

Gary reviewed the committee's previous decisions as he understood them. If the Amalgamated sent representatives directly to U.S. Steel they would be told that the corporation was only a stockholder in its subsidiaries and would not undertake to "instruct or interfere" in their local affairs. The question now before the committee was whether it should authorize subsidiary presidents in fact to deal with the union so long as under no circumstances they recognized extension of the union into nonunion facilities. At previous sessions of the committee it had been suggested that subsidiary presidents, when demands were made by the union, should ask for a day's delay, immediately bring the matter to the executive committee for instructions, and reply to the Amalgamated the next day. Schwab doubted that anyone would be fooled by that procedure.

Gary's concern, however, was that the executive committee retain full control and be free to alter policy as events dictated. He was far less sure than Schwab as to what the Amalgamated would do. If the union pursued its stated policy of treating the subsidiaries as a unit and were rebuffed in its demands, it probably would strike. Schwab replied that a strike was but "a remote possibility."

Roberts agreed with Gary that U.S. Steel should refer the Amalgamated to the subsidiary companies for discussions, but opposed the corporation making any "hard and fast rules" until the union spelled out its demands. Meanwhile, he believed subsidiary officers could consult with their "local" (subsidiary) boards of directors. Converse went further. He recommended that Schwab tell the subsidiary presidents that U.S. Steel was a "large financial institution" and expected them to handle their own labor problems "just as if the United States Steel Corporation did not exist." He was certain they would "be very careful and not get into trouble." Schwab, Steele, and Reid all gave "unqualified approval" to the suggestion.

After further discussion, Steele introduced a resolution far more moderate than the tone of the debate suggested. Although the committee "unalterably opposed any extension of union labor and advise[d] subsidiary companies to take a firm position," "great care" was to be taken to prevent any trouble. If difficulties arose, the subsidiaries were to report it "promptly" to the corporation. Schwab's rhetoric seemingly had carried the debates; in the end, however, the committee decided to keep matters in its own hands and to be cautious and flexible as Gary proposed. "Temporizing," adopted as policy in April, remained intact.[10]

Perhaps encouraged by the tone of the debate on June 17, Schwab on July 1 once more made a play to free himself and the subsidiary officers from the heavy hand of the executive committee. Schwab gave each member a copy of a "proposed plan of operation." The president of U.S. Steel and four of his associates were to have complete charge of the affairs of the subsidiary companies. In the heated discussion that followed, Gary observed that the proposal set up a mode of procedure not contemplated in the bylaws of the corporation. What, precisely, would be the function of the executive committee under the new scheme? Schwab answered that it was impracticable for him to take up "ordinary matters of detail" with the committee. When Gary responded that he believed "the head of the financial house" (Morgan) had a different view as to how the corporation was to be run, Schwab repeated his charge that the existing system was "impracticable." [11]

At a session the next day, a member of the committee introduced a resolution quoting those parts of the bylaws that spelled out the duties of the executive committee. The resolution noted that the president, believing the corporation could not be operated successfully in the prescribed manner, thought it "necessary" to "disregard" those procedures. The entire matter, including all differences of opinion on the subject among members of the committee, was to be submitted to the full board of directors for action. In the meantime, it was the sense of the meeting that unless the bylaws were amended, affairs of the corporation were to remain subject to the control of the executive committee. Schwab argued that he only wished to conduct the affairs of U.S. Steel "as similar organizations have been conducted in the past." Some members sympathized but were unwilling to support Schwab unless and until the bylaws were changed. Gary had no wish at this point for a test of power between himself and Schwab before the board of directors. He also knew that U.S. Steel badly needed Schwab's expertise in steelmaking. Content with the substance of authority, and seeing no advantage in humiliating or repudiating Schwab, he tried to head off the resolution. It gave the appearance of a clash of personalities, he argued. He also thought it unwise for the committee to take such matters to the board of directors where it "might lead us into trouble." Direct consultation with Morgan, who at the moment was abroad, seemed to him the best approach. The committee, however, having finally reached a decision, refused to be stayed and adopted the resolution. It would continue its supervision of the subsidiary companies. Schwab, "simply heartbroken" at the rebuff, for the moment considered resigning. [12]

After July 2, a subdued Schwab abandoned his efforts to win independence from the executive committee. The lawyer-bankers, however, no longer trusted him. On July 11, Roberts introduced a resolution that the chairman (Gary) was to represent and to exercise all the powers of the executive committee between its sessions which, with the coming of summer, would be more irregular. Gary "had no objection," he said, "provided it was agreeable to the President." When Schwab declared that the resolution was "perfectly satisfactory" to him, the committee adopted it.[13] Gary's victory over Schwab and the steelmakers for the moment was complete.

Even as the executive committee was voting against Schwab's plan to escape their supervision, problems with the Amalgamated came to a head. T. J. Shaffer, president of the union, by July 1 had approached several of the subsidiary companies, demanding that they accept a wage scale for both their union and nonunion mills. The companies agreed to the scale but only for those shops that were already unionized. Shaffer gave one of the subsidiaries, American Tin Plate, until July 8 to accede to the demands or face a strike. The executive committee noted that a strike apparently could be avoided by ceding only three mills to the union. Gary thought this "a pretty good way out of a bad thing." It was, after all, "the very worst time of the very worst year to have any trouble." Reid and Converse agreed. Edenborn and Roberts preferred not to "give way one inch." As Edenborn saw it, "a concern operating with a union is pretty badly handicapped." Roberts had cooled down by the next day and concluded that the time was not ripe for a showdown.[14]

Meanwhile, Schwab had met with Morgan and reported that while the financier would support any steps taken, he clearly wanted the affair settled. The executive committee's plan of having subsidiary companies, under its guidance, deal with the union one by one, and that no nonunion shop be surrendered to the union, began to erode. On July 6 it voted to allow three subsidiaries on their own to negotiate simultaneously with the Amalgamated. When the superintendent of a sheet mill discharged a dozen employees for attempting to organize a union lodge, the committee reluctantly ordered the men reinstated.[15] Meanwhile, the subsidiary companys' negotiators ceded six, not three, mills to the union. When the coporation rejected that concession, the Amalgamated struck all three subsidiaries.

Although the strike received only tepid support from the rank and file and less from the public, the corporation continued to seek a compromise settlement. Eventually Schwab, and then Morgan himself, conferred with union officials and offered to surrender some mills that had not before been unionized. The union held out for total victory rather than settling for the compromises that it easily could have had. At the end of July it struck all U.S. Steel subsidiaries and lost. Only a skeleton of the once-powerful Amalgamated hung on in a few properties of U.S. Steel until they were routed in a brief strike in 1909.

Events in July made hash of Gary's tactics for handling the Amalgamated. But the so-called practical steelmen had not done much better. Gary and Schwab had toured the subsidiary companies together in May and reached quite different conclusions. Schwab confidently, and inaccurately, had predicted that a strike was unlikely. Gary had observed that if the Amalgamated carried out its threat to deal with the subsidiaries as a unit, "a big strike" would come, at least in the unionized plants. And, while Gary's "temporizing" failed, and the steelmen's tougher stance towards labor appeared to be vindicated, Schwab's insistance that subsidiary officers could best handle the labor problem had produced unacceptable results. Fortunately, from the view of both factions, the affair was over.[16]

* * *

Schwab's defeats in the executive committee on July 1 and 11 marked the beginning of his fall from the presidency of U.S. Steel. Unfavorable newspaper publicity, Schwab's flair for high living, and his entanglement in outside business ventures also contributed to his downfall. Coming to sudden wealth, Schwab, not unlike several other Carnegie proteges, did not always behave prudently. Amid newspaper speculation that U.S. Steel paid him a salary of $1 million a year (he actually received $100,000 in salary and a bonus that in 1901 also amounted to $100,000), he had begun construction of a mansion in New York City modeled after a French chateau. Slowly the four-storied building—more a luxury hotel than a home—took shape. It boasted its own electric generator, had a 116 foot tower, six elevators, ninety bedrooms interconnected by telephone, a wine cellar, swimming pool, bowling alley, and gymnasium among its many features. As his biographer puts it, "Riverside" was "the last, the largest, and the most lavish, full-block mansion ever to be constructed in New York City." [17]

At the close of his first trying year in office, Schwab decided to take a vacation in France. There he fell in with a wealthy international set. As the press reported it, Schwab, with his newfound friends, dashed from casino to casino on the Riviera in a flashy big automobile that he bought in Paris. One night, lucky at the roulette wheel, he reportedly broke the bank at Monte Carlo. The story, though untrue, shocked Carnegie. It was "as if a son had disgraced the family," he wrote Morgan, and called for Schwab's removal as president. The financier, somewhat more tolerant, refused to accept Schwab's resignation when it was offered.[18] Nevertheless, in the months that followed, Morgan increasingly supported Judge Gary's policies in the corporation. At one point, when Gary threatened to resign because of differences with Schwab, Morgan threw his full support behind the chairman. "Now you remain where you are," Morgan told Gary, "and, from this time on, when you want me to do anything or say anything, all you have to do is tell me." [19]

Meanwhile, Schwab involved himself in a number of other businesses including the International Nickel Company, American Steel Foundries, the United States Shipbuilding Company, and Bethlehem Steel. U.S. Shipbuilding, founded by the merger of several firms in 1902, by mid-1903 was in financial straits. As the firm sank into bankruptcy, Schwab alone emerged financially unscathed. Lawsuits followed. The resulting publicity, which portrayed the president of U.S. Steel as a speculator, became more of a burden than Morgan or the corporation was willing to bear. Schwab resigned on August 4, 1903, to become president of Bethlehem Steel, in which he had acquired extensive interests.[20]

Corey, now thirty-seven and president of Carnegie Steel, replaced Schwab. Regarded as a model of propriety, he was described as "stern, chilly and aggressive," with "no fads except a fondness for home." [21] The change in no way injured Dickson's career; he continued on as second vice-president of U.S. Steel, the position to which he had been elected in 1902.

* * *

Once the Amalgamated was defeated and Schwab's independence curbed, the officials of U.S. Steel turned to their first major venture in welfare capitalism. They adopted "profit sharing" in the form of an employee bonus and stock purchase plan. The rationale for the scheme was that it would give employees an interest in the firm and incentive to work more efficiently. Although not widely practiced in America at the time, the idea was not new. Even some of U.S. Steel's subsidiaries, including Carnegie Steel, had profit-sharing programs in operation.

The plan adopted by U.S. Steel, however, was more comprehensive than most and was designed to cover an unusually large labor force. George W. Perkins, a Morgan partner elected to U.S. Steel's board of directors in November 1901 and thereafter chairman of the board's finance committee, was its chief architect. The bonus portion of the plan applied only to managers of U.S. Steel and the subsidiary companies and not to wage earners because their wages were so low that their share would have amounted to perhaps no more than a dollar or two a year—too little to have much impact on their attitude toward their job or the corporation. The plan called for a percentage of all corporation earnings beyond expenses, interest on the bonded debt, and a regular dividend to shareholders be set aside for division among the managers.

The stock purchase plan, however, was open to all employees and offered a means for sharing in U.S. Steel's profits. Each year, employees were invited to buy shares of stock (at slightly less than market price) on an installment plan. To encourage purchases and to induce workmen to remain with U.S. Steel, subscribers received the regular dividend payments on the shares they were buying. Also, for a five-year period (providing they remained at U.S. Steel), they were given an annual five-dollar bonus on each share. To further sweeten the plan, the corporation continued to pay dividends and bonuses on shares whose subscribers withdrew from the plan, and the resulting fund was divided up among those still in the program at the end of five years.

Perkins spoke of his plan as embracing everyone "from the President to the man with pick and shovel." Critics have charged, however, that he and the corporation intended the plan only for a relatively small but valuable group of employees—the skilled workers—who also happened to be the most vulnerable to unionization. Perkins's most recent biographer found no evidence of such intent, but conceded that, as the plan worked out, the chief beneficiaries were indeed the skilled workmen. Some of the planning by Perkins in drawing up the scheme makes it difficult to believe that he was much surprised when the great mass of common laborers failed to participate. He began by dividing U.S. Steel's 168,000 employees into six categories. The top group, Class A, was made up of the fifteen managers who earned salaries in excess of $20,000 per year. At the other extreme, nearly three-quarters of the entire labor force, with incomes below $800 per year, were grouped in Class F. The average income level of this group fell between $500 and $600 per year. Surely Perkins realized that few workmen in that wage category could afford shares of stock even on the installment plan, given the fact that shares offered in January 1903, when the plan began, sold for $82.50 each. That 10 percent of the employees in Class F signed up to buy shares is far more surprising than that virtually all of them dropped out of the program within a month or two, or that 90 percent of them avoided the scheme altogether.

In its first year the program drew 13,000 subscribers from the higher paid categories of workmen. In testifying before the Stanley Committee of the House of Representatives, which was investigating U.S. Steel in 1911, Perkins noted that 30,000 (15 percent) of the employees of the corporation were subscribers in the stock purchase plan. Nearly 85 percent, of course, were not involved. Although the total number of employees in the program reached a total of 42,258 in 1918, given the swollen employment rolls by that date, the rate of participation remained between 15 and 16 percent.[22] In explaining the advantages of the plan to the Stanley Committee, Perkins made no attempt to hide its usefulness against unionization. Stock ownership, he explained, made the men "more competent to pass intelligently on the questions that might be involved in a contemplated strike." As stockholders they would "regard the question from the standpoint of a partner rather than from the standpoint of a mere hireling." When asked if higher wages might not be more important than dividends to the workers, Perkins replied that it wasn't the "stock income" that mattered so much as that the employee was becoming a capitalist; "it is the investment he is making." [23]

Although a few of the better paid employees could afford to buy stock and did receive a good return on their investment, Perkins's plan came nowhere near involving the great mass of workmen. Far from benefiting them, the plan served chiefly as a public relations gimmick of the corporation whenever more meaningful reforms were suggested.

* * *

No record has been preserved of what role, if any, Dickson played in the labor polices of U.S. Steel before 1906, even in his own papers. As assistant to President Schwab at Carnegie Steel, Dickson's duties had involved the purchasing of scrap and pig iron and keeping the president "informed as to current operations, making monthly comparative statements of the production of every furnace and mill." Transferred with Schwab to U.S. Steel, Dickson coordinated the various activities of the constituent companies, meeting regularly with traffic managers, sales managers, and purchasing agents. The presidents of the subsidiary companies met each month in New York. Gary usually presided. In his absence, Schwab took over and, on a few occasions, Dickson.[24] Given his position as Schwab's assistant, his relationship with the steelmaker faction, and his own distrust of unions, Dickson probably sided with the steelmakers on how best to meet the threat of the Amalgamated in 1901. He, of course, had no part in the stock purchase plan; that was Perkins' special project. If Dickson's views in 1902 and 1903 coincided with his later views, however, he heartily supported the program.

After 1906, Dickson became increasingly involved in labor reform at U.S. Steel. The problem of industrial accidents and deaths became a major concern of the corporation in May 1906 when the *Chicago Tribune* ran a series of articles criticizing the corporation's hospital at its South Chicago plant. The deaths of men accidentally killed in the mills, the articles alleged, went unreported to city officials; men injured in the plant were duped at the hospital into signing away their rights to sue the company for their injuries. That same month, casualty managers of U.S. Steel's subsidiary companies met together for the first time to discuss the overall problem of accidents.[25] Within the year, Gary was considering the establishment of a general accident relief fund for all corporation employees. Dickson wrote Charles L. Taylor, chairman of the Carnegie Relief Fund, for data on accidental deaths in the Carnegie Division of U.S. Steel during 1906. Exclusive of deaths due to direct negligence on the part of the victims or where the deceased was "not a contributor to the support of his family" or where the death did not otherwise fall within the scope of the fund, 178 men had died from industrial accidents.

In forwarding this information to Gary, Dickson offered to collect similar data from the other subsidiaries. The inquiry was made, and Dickson reported a total of 405 fatalities in all divisions of U.S. Steel during 1906.[26] Dickson proposed that whether or not the corporation decided to adopt a "general system of relief," some "competent man or men" should be appointed to make "a thorough inspection of every works, mine, etc. with a view to making such recommendations to the Presidents as to additional safeguards as may seem practicable." This proposal, Dickson later claimed, was the "genesis of the 'Safety Dept.' of the U.S. Steel Corporation." [27]

Dickson's proposal was already under discussion when William Hard published "Making Steel and Killing Men" in *Everybody's* magazine in November 1907. Hard's public disclosure that forty-six men had been killed and perhaps two thousand others "merely burned, crushed, maimed or disabled" at the South Chicago plant during 1906, no doubt induced Gary to act quickly. By April 1908, the Safety Committee had been established. In October 1910, Dickson reported that fifty corporation employees were devoting substantial time to the safety campaign and that the project was costing the corporation $400,000 per year. By 1916 the cost of the program rose to $750,000. Chiefly through education (short talks to the men and safety messages posted on bulletin boards in the shops and enclosed in pay envelopes) and competitive interdepartmental safety contests, U.S. Steel reduced serious accidents by 43 percent within four years.[28] One corporation official boasted that what had begun as "a species of self defence" had "broadened out into more humanitarian lines" until it was "being taken up on a scale that could not have been dreamed of in this country a few years ago." U.S. Steel's safety program became a model for similar programs in many other firms across the nation.[29]

The corporation's voluntary accident relief program evolved directly from its safety drive. In December 1909, Charles MacVeagh, general solicitor of U.S. Steel and head of the five-man Committee of Safety, wrote Dickson with regard to compensation for injured employees. A number of European countries, he noted, had adopted the idea that "the cost of industrial accidents should be made a part of the cost of production and that the men injured and the families of men killed should be given compensation or relief irrespective of employer negligence." Several American states, he pointed out, were considering similar legislation or had appointed commissions to study the matter. Meanwhile, the common law defenses of employers against injury suits (assumption of risk or negligence on the part of the injured person, or contributory negligence by a fellow workman) in many states had been weakened or done away with. Decisions in such cases were increasingly being left to juries to decide. "This makes it more and more necessary to avoid litigation," MacVeagh warned, and U.S. Steel should, "to the greatest possible extent," rely on "amicable settlements."

The plan advanced by MacVeagh offered injured workmen certainty of relief without having to apply to the company for charity or to threaten litigation and settlements that the recipients would not have to share with attorneys. At the same time the plan protected the corporation from lawsuits and heavy judgments by juries. The scheme was wholly voluntary, and the corporation financed it without cost to all employees. Workmen were covered as long as they remained with the corporation. To receive payments, a victim's injuries had to be certified by a company physician, and the employee had to waive the right to sue the corporation. Adopted in January 1910, the plan provided for free medical treatment of injured employees, guaranteed the injured employee a third or more of his usual daily wage during recuperation, provided between six and eighteen months' wages for permanent injury, and offered death benefits of eighteen month's wages to the victim's survivors, plus additional sums according to length of service and the number of minor children left unsupported.[30]

As the accident relief scheme was going into effect, Charles Taylor and Dickson began urging adoption of a general pension plan for superannuated employees of U.S. Steel. Carnegie, upon his retirement, had set aside $4 million to provide old age pensions for his employees; this fund covered about a third of U.S. Steel's 200,000 employees. If the corporation were to add $8 million dollars to Carnegie's fund, the 130,000 employees not covered could be provided for. "I have been taking every favorable opportunity to urge upon Mr. Corey [Schwab's successor as president of U.S. Steel] and Judge Gary the adoption of some definite Pension plan which can be announced to the workmen," Dickson wrote Taylor on March 2, 1910. Within a few weeks Carnegie, the finance committee of U.S. Steel, and others concerned had agreed to the proposal, and the pension plan was extended to all corporation employees in January 1911.[31]

Dickson could well take satisfaction in the labor reforms that he helped shape at U.S. Steel—accident relief, safety, and pensions for retired workers. Although paternalistic and not without benefit to the corporation, they did directly improve the lot of workers in more than trivial ways. This was less true of two other aspects of the corporation's welfare capitalist program that developed before World War I. One, "community health work," included "sanitation, medical services, restaurants and housing." [32] The corporation regarded the services provided under this category as a form of benevolence; most self-respecting employers would have regarded many of them as ordinary housekeeping expenses: providing employees with cool drinking water, clean eating places, decent toilet facilities, and the like. U.S. Steel's assistant general solicitor, addressing the American Iron & Steel Institute in 1912, listed the installation of a sewage disposal plant among the corporation's projects at one mill. The "drainage of stagnant water," he also noted, had "become another source of concern. The enforcement of cleanliness and order, fencing, painting, cutting weeds, and collecting garbage," were among "the details" that had come "under the consideration even of the presidents of subsidiary companies." [33]

Of greater value to workers and their families were company housing, company towns, and loans to employees for buying homes. These programs improved the living conditions of workmen fortunate enough to be located in the newer, model communities as opposed to those stuck away in older and more slumlike company housing. Unfortunately, housing projects had the great disadvantage of making workmen doubly dependent on the corporation. It was now not only their employer but also their landlord or mortgage holder. U.S. Steel, it should be noted, was not above exploiting this dual relationship in its struggle to avoid unionization of its men.[34]

More amorphous was the portion of U.S. Steel's program labeled "welfare." This catchall category included corporation contributions to churches, hospitals, libraries, schools, charity, employee picnics, and athletic facilities, to name but a few. So general was this heading that one year it may even have included a contribution by the corporation to the Protective Tariff League. Over the years, "community health work" and "welfare" took up more and more of the corporation's total expenditures on welfare capitalism. In 1912, 22 percent went to these two categories. By 1918 they received 54 percent of the funds. By contrast, accident relief, safety, and pension expenditure (above and beyond that covered by funded investment), fell from 60 percent to 35 percent of the total.[35]

* * *

Dickson launched the second phase of labor-reform-from-above at U.S. Steel in 1907 when he undertook to reduce the hours of labor put in by steelworkers. He began modestly, seeking first the abolition of all unnecessary Sunday labor at all U.S. Steel properties. He thought a shorter workday was also important, but he knew from personal experience that both the long day and the seven-day workweek were firmly entrenched and thought it best "not to extend the battle line too far." [36]

When he presided at a meeting of the subsidiary presidents in early April, Dickson somehow secured an "informal action" against unnecessary Sunday work in the mills. Converting that into a draft resolution, he asked President Corey to bring it to the attention of the finance committee for its "official recommendation." Dickson's resolution called for reducing Sunday labor "to the minimum." Except for repair work that could not be done while operating, such labor was to be suspended at "all steel works, rolling mills, shops, mines, quarries and docks." Construction work and the loading and unloading of raw materials were also banned. It was understood, of course, that the recommendation could not apply everywhere—and most notably not in blast furnace operations.

Dickson's draft included a rationale. The resolution was made not only to express the "most earnest conviction" of the committee, "nor alone in deference to public opinion," but for "the benefit of the immense number of employees affected—and in the belief that the day of rest will make them better citizens and more efficient workmen, and that instead of loss, the Corporation will eventually be benefited." After tightening up Dickson's phraseology, eliminating his explanation of motives, and rewording the conclusion to read that it was "desirable that the spirit of the recommendation be observed to the fullest extent within reason," the finance committee on April 23 adopted the measure.

Highly pleased with his work, Dickson sent copies to the subsidiary presidents. As he later observed, he was still innocent enough to assume that a finance committee resolution "would mean what it said." [37] The subsidiary presidents were not so naive. They complied with the ban briefly in 1907 while conditions were depressed in the industry. Only blast furnace men continued working seven days a week, and they found it relatively easy to get a day off by asking. Other firms, if aware of U.S. Steel's policy, did not abandon Sunday labor. When, after 1909, business revived, the subsidiary presidents were determined to meet production schedules at minimum costs and simply "forgot" the resolution. By the end of 1909 "seven day operation of steel works and rolling mills was common;" indeed, "practically every plant started operations for the week on Sunday afternoon." [38]

Meanwhile, outside pressures for labor reforms were starting to build. Hard's muckraking article first drew attention to conditions in the steel mills. In 1907 and 1908, a team of young scholars under the direction of Paul U. Kellogg, and financed by the Russell Sage Foundation, began the celebrated Pittsburgh Survey. Their insistent probing for the roots of Pittsburgh's many problems in due course led them to the steel industry and its inhumane labor policies. Full issues of *Charities and the Commons* (soon to be renamed *Survey*) were given over to their findings in January, February, and March 1909.

Not long after the February issue, Senator Robert LaFollette condemned the steel industry for working men eighty-four hours a week and sometimes for twenty-four hours at a stretch. To rebut the Senator, Gary asked Dickson to prepare a report on hours worked by U.S. Steel employees. Dickson discovered that despite the 1907 resolution against all unnecessary Sunday labor, the number of men actually working on Sundays in the various plants ranged from 1 percent of the work force in some to 80 percent in others.[39] Two of the largest subsidiaries, Illinois Steel at Gary and Carnegie Steel in Pittsburgh, were among the chief offenders. The Universal Portland Cement Company of Chicago, a subsidiary that turned out no steel whatever, regularly operated its plants seven days a week. Under question by Dickson, Edward M. Hagar, the firm's president, defended the practice on several grounds. He estimated that to close down on Sundays—instead of only on Christmas, Easter, and the Fourth of July, as was the practice—would cause output to drop one-sixth and net profits between $400,000 and $600,000 a year. Since kilns could not be allowed to cool without damage, Hagar declared, Universal, like its competitors, ran continuously.

As for ethical considerations, Hagar believed that the existing system was preferable to Sunday shutdowns. To restrict Sunday labor to necessary repairs only would necessitate work by "the very class of men that can afford to take the day off and would use Sunday properly," whereas under the existing system those men were usually off. On the other hand, the mass of unskilled workmen, though free to take off a day whenever they asked, preferred to work seven days a week to maintain their incomes. Unless they were given seven days' pay for six days' work, enforced Sunday layoffs would injure them "financially and morally" and give them less money for educating their children. With Sundays off, the majority would only squander their time and money in saloons. Hagar cited as evidence the difficulty on the Monday after Easter when many men came in "unfit for work." [40]

Dickson reported to Corey that if steel companies could shut down open hearth furnaces, soaking pits, and heating furnaces on Sundays, Hagar's contention that cement kilns had to be run seemed improbable. As for Hagar's arguments concerning men and their habits, Dickson doubted they would be of much use "if it ever became necessary to make a public defense of his practice of operating every day in the year except the three mentioned." Dickson urged Corey to take appropriate action to enforce the 1907 resolution.[41]

By May, Dickson had read the Pittsburgh Survey and sent a copy to Corey with all passages commenting on long hours or Sunday labor carefully marked. Dickson observed that Abraham Lincoln, as a youth, had traveled by flatboat down the Mississippi, had witnessed the abuse of slaves, and had resolved to destroy slavery if ever given the chance. Dickson's own feelings about Sunday labor, he said, had similar origins. The "hardship of a seven day week made a profound impression on me when working as a boy of fifteen in the Homestead mills." He doubted that anything—even higher wages—would "so commend the Corporation to the public and to the workmen" as taking an advanced stand against the seven-day workweek and firmly maintaining it. The corporation's dominant position in the industry, he predicted, would "speedily compel" competing firms to follow its example.[42]

In January 1910, Dickson called Corey's attention to growing attacks on the corporation from labor leaders. Their criticism, he noted, stressed the issue of Sunday labor—the "one weak point in our armor." If the public became aroused by revelations such as the Pittsburgh Survey, he had no doubt that the company's "present methods of operating" would be condemned. A recent issue of *Survey*, he pointed out, admitted the need for continuous operations in some phases of steelmaking but argued that this in no way justified using the same men all seven days. By staggering crews, all workmen could be given one day off in seven. Dickson endorsed that idea fully and proposed that U.S. Steel and its subsidiaries arrange operations so that no man would work for more than six consecutive days. As for the argument that workmen were already at liberty to take days off, Dickson observed that both he and Corey knew "from our own experience in the mills" that while true in theory, it was not true in practice. The solution, Dickson said, was to adopt a system that compelled absence from work every seventh day except in cases of emergency.[43]

By 1910, Dickson's discreet campaign against unnecessary labor on Sundays had become a one-man crusade against the seven-day workweek. Elevated to the first vice-presidency of U.S. Steel in 1909, Dickson steadily argued for the reform both among his associates at 71 Broadway and at monthly meetings with the subsidiary presidents. When nearly three years of polite reminders and prodding about the 1907 resolution came to nothing, Dickson decided on a showdown. At a meeting of the subsidiary presidents on February 25, 1910, with Judge Gary presiding, he read aloud his recent letter to President Corey, and invited discussion.[44] E. J. Buffington, president of Illinois Steel, observed that it required nearly as many men to keep open hearth furnaces heated while idle on Sundays as it did to operate them for production. According to his superintendents, there was "very little gained by not operating." The majority of workers, moreover, sought the "opportunity for seven days' work" so as to increase their incomes. Dickson asked Buffington if he thought he got as much work out of men who labored 360 days a year as from those who worked only 300. A six-day workweek, Buffington replied, would require one-seventh more capacity in every department. Dickson admitted this, but argued that even in terms of dollars and cents he had no doubt that a reduced workweek would prove more efficient. He asked Buffington if he really believed that any large percentage of his men preferred to work seven days a week. Buffington conceded that they would not if they could earn in six days what they presently earned in seven. A change of that scope, however, could be accomplished only by "revolutionizing our methods."

The discussion clearly distressed Judge Gary. In an apparent attempt to sidetrack the matter, he suggested that Dickson appoint a committee of "five of our best men" to investigate and report within three months on how many employees were working seven days a week. Dickson was prepared, having already collected that information from the subsidiary presidents. "Would you like a statement to go out that 50% of the men, in some instances, work seven days a week?"

Judge Gary then read the 1907 resolution which called for an end to all unnecessary Sunday labor. Dickson declared that the "spirit" of the resolution had not been observed. When spokesmen for the largest subsidiaries, Dinkey of Carnegie Steel and Buffington of Illinois Steel, denied the charge, Dickson pointed out that the new mills at Gary, Indiana, "were built with 80% of the men working on Sunday." Did that not "violate the spirit of that resolution?" he asked. Gary objected to Dickson putting Buffington "on trial." It was "only fair to say we all knew it." "That shows that I charge in this matter," Dickson snapped back, "—insincerity."

Gary inquired as to the number of hours that U.S. Steel employees actually worked. "In the steel works, the majority work 72 hours," Dickson said; "in the blast furnaces, 84 hours per week." Dinkey added that mechanics and day laborers ordinarily worked only sixty hours. The Judge was perplexed. "I do not know what to say. . . . I do not like to make the flat statement that there is much of a moral question involved." As he saw it, the problem was that "from the standpoint of making the most money for the Corporation, there might be one answer, and from the standpoint of doing the right thing, taking into account everybody, there might be another." He tried to escape the dilemma with an appeal to piety: "We say that if you are sure of your morals, good policy follows."

Dickson, however, insisted that his own arguments stemmed from an economic, not a moral point of view. The seven-day workweek brutalized men and decreased their efficiency. But, Gary persisted, "are we doing the laborer an injury?" "Absolutely," Dickson declared. "If we have a repetition of 1892 [the Homestead Strike], . . . they will turn and rend you; the very men you brutalize and deprive of the opportunity to live normal human lives."

Dinkey took "the other side entirely." He did not believe that "a single individual in our concern" had been brutalized. "There are men suitable for iron work and men unsuitable. I know of very few men in our concern whose efficiency has been seriously, or to a small extent interfered with." "Would you go before the public and advocate seven days a week?" Dickson asked. "That is different," Gary insisted.

When the Judge asked why Illinois Steel, for example, found it necessary to use 50 percent of their men on Sundays, he was told that it was due to "most extraordinary pressure from their order book." Under the pressure of orders and the desire of the workmen for more income, Buffington noted, his mills had sometimes operated on Sundays without his prior knowledge or consent. W. P. Palmer of the American Wire Company reported that his superintendents were "making an honest endeavor to reduce Sunday work to a minimum," but he questioned whether his firm could run without Sunday work "and make the returns expected of it."

Profitmaking clearly was at the crux of the matter, but in concluding the discussion Gary completely evaded that issue. "I understand you, Mr. Dickson, to put this on the ground that we are actually injuring the men, therefore ourselves. I am not satisfied we are injuring our own men; whether we are injuring the morals of our men, that is another question."

The "showdown" disappointed Dickson and widened the fast-developing gulf between himself and his colleagues. "My fellow officials, men with whom I had been associated from boyhood," he observed, "first ridiculed my suggestion as impractical idealism and then denounced me as a disturber of an ideal social order." Far from curbing his determination, however, the rebuffs served to arouse his "fighting blood." [45] A week after the clash, Dickson demanded from Dinkey an explanation—if his plants were observing the rule against Sunday labor, why their combined Saturday-Sunday output of steel exceeded the usual daily tonnage by more than a third. Dinkey's answer was forthright: it was done "by charging earlier and *adding a little to the necessary help which is regularly on duty* taking care of the furnaces and tending gas." [46] Clearly the 1907 resolution held little dread for him.

Dickson forwarded Dinkey's reply to Corey. He also sent a summary of the daily output of open hearth steel ingots at Dinkey's and Buffington's companies for the first two weeks of March 1910. The data showed that the combined Saturday-Sunday output at Carnegie Steel for the two weeks amounted to 160 and 153 percent respectively of the average daily output for the preceding five days. The corresponding figures for Buffington's South Chicago and Gary plants were 174 and 187 percent and 182 and 203 percent. No disciplinary action was taken against either president.[47]

Sure of his cause and sensing that the time was ripe for reform, Dickson pressed on. In a search for allies, he secretly invited Paul U. Kellogg of the Pittsburgh Survey to his office. "I want to tell you fellows to keep up your pounding from the outside," he told the surprised reformer. "There are some of us here at work with picks and crowbars on the inside and we'll have that old wall down together. Yours is the first bit of encouragement we have had that the public knows or cares about the long day and the long week in steel." [48]

As would be true of his later victories, Dickson's success on the Sunday labor issue was made possible by unexpected outside pressures. During the first week of February 1910, the unorganized employees of Bethlehem Steel quit work because of excessively long hours, unnecessary Sunday work, and low wages. The strike dragged on into March. When bloodshed occurred, Congressman A. Mitchell Palmer of Pennsylvania induced the federal Bureau of Labor Statistics to begin an investigation. On March 16, both strikers and Bethlehem officials began to give testimony. When it was revealed that 51 percent of the men worked twelve hours a day, 43 percent worked seven days a week part of the time, 29 percent worked seven days a week regularly, and 46 percent earned less than $2 a day, President Schwab tarred the industry at large by insisting that conditions at Bethlehem were not unique; they prevailed in most iron and steel mills.[49]

In the midst of the Bethlehem Strike, Dickson dropped a bombshell at 71 Broadway. In a forceful letter addressed to Corey, his immediate superior, but intended for Gary, Dickson reviewed his efforts since 1907 to end the Sunday labor abuse. He charged that in the subsidiary companies "a vast amount of work" was done on Sundays for which there was "no excuse except the desire of those immediately in charge to make tonnage records and increase profits" without regard for the "natural rights of the employees to have sufficient free time to live a normal human life." Citing acknowledged violations of the 1907 resolution (such as Dinkey's) and the corporation's failure to punish them, Dickson suggested that "those responsible for the policy of our companies are deliberately electing to coin profits out of the very lives and souls of the workmen." The power to effect a change, he declared, rested squarely on Corey and Gary. Because of the "spirit of indifference, if not actual hostility" displayed at the showdown meeting on February 25, Dickson concluded he had "exhausted all the means in my power as an officer of the Corporation" and nothing remained but "to place on record this formal protest against a continuance of the conditions which I have described." [50]

Gary read the letter and wrote on it that he was "personally in general accord with the views expressed," both by the letter and by the finance committee resolution against Sunday labor "which I drew." Corey, too, endorsed Dickson's letter. Both men, in fact, totally opposed the vice-president's course. In Dickson's opinion, Corey was a "fair fighter." Gary was "evasive and insincere." Corey reported to Dickson that Gary had been incensed by the letter, labeling it "a gross impertinence." [51]

Because of the investigation at Bethlehem then in progress, Gary and Corey were in no position to resist Dickson's demands, and Dickson knew it. When Corey showed the vice-president a draft of a proposed letter to the subsidiary presidents on the Sunday labor question, Dickson called it weak. Corey's past actions in the matter, Dickson said, proved to him and the subsidiary presidents that the president was insincere. Unless "immediate and drastic action" were taken to force compliance with the resolution, Dickson threatened to resign at once and "try what I could do from the outside." That same day Congress passed a resolution ordering an inquiry into the impact of the Bethlehem Strike on pending government contracts.[52]

The next day Dickson met with Gary. If the letter of the fifteenth seemed impertinent to Gary, Dickson's verbal remarks on the eighteenth must have appeared insolent. It was clear, Dickson said, that the subsidiary presidents doubted both Corey's and Gary's sincerity regarding Sunday work. Dickson questioned whether Corey's letter would be believed and asked if Gary would supplement it with telegrams to the presidents "making it very clear . . . that both the spirit and letter of the Resolution must be observed." In urging the matter on the chairman, Dickson declared that he was offering Gary "the opportunity of a lifetime" to place his name "among the benefactors of your race." Corey had had his opportunity and failed. Would Gary "now do for the steel workers what Lincoln did for the Negro?" Dickson ended his carefully prepared remarks with a threatening question: if the presidents were permitted to operate as usual on the coming Sunday, would he not be "justified in concluding" that he "need look for no real assistance in this matter within the Corporation?"[53]

Gary yielded and with Dickson's assistance prepared a telegram to the subsidiary presidents:

> Mr. Corey, Mr. Dickson and I have lately given much serious thought to the subject matter of Resolution passed by Finance Committee April 23rd 1907, concerning Sunday or Seventh Day Labor. Mr. Corey has written you on the subject within a day or two. The object of this telegram is to say that all of us expect and insist that hereafter the spirit of the Resolution will be observed and carried into effect. There should and must be no unnecessary deviation without first taking up the question with our Finance Committee and asking for a change of the views of the Committee, which probably will not under any circumstances be secured. I emphasize the fact that there should be at least 24 continuous hours interval during each week in the production of ingots. Please acknowledge.[54]

Having given in, Gary decided to reap as much public good will from the order as possible. On March 21 he called to his office Paul U. Kellogg, John A. Fitch (who had written much of the Pittsburgh Survey's report on conditions in the steel industry), and Charles M. Cabot (a Boston broker and U.S. Steel stockholder who was a critic of the corporation's labor policies). With Dickson at his side, Gary read the telegram aloud and handed it to Dickson to dispatch to the subsidiary presidents. Despite his earlier contact with Kellogg, Dickson "didn't bat an eye of recognition." By the end of the year a congressional investigation reported that fewer than 5 percent of U.S. Steel's employees still worked seven days a week.[55]

Despite the fact that his agitation to shorten hours of labor increasingly irritated Judge Gary and other officials of the corporation,

Dickson persisted. About the time that the Bethlehem Strike Report was made public in May, Dickson wrote Gary that in "the near future" the corporation would have to face the twelve-hour question. "Any thing in the nature of compulsory legislation on this subject would present a serious problem to our Companies. Would it not," he asked, "have a tendency to prevent too radical action if we recognized the hardship of an eight-four (84) hour week, even in the processes which are necessarily continuous, and voluntarily adopted a system whereby no man would be on duty more than six days consecutively?" [56]

Gary, who liked to think of himself as an enlightened employer and a friend and benefactor of labor, resented Dickson's intrusions in the area. To make matters worse, the vice-president had become a thorn in Gary's side on other questions as well. Over the years Dickson had led opposition within the coporation to the chairman's commercial policy. And recently he had supported yet another attempt to preserve the autonomy of the presidents of U.S. Steel and the subsidiary companies from Gary's growing power. From the Judge's point of view, Dickson was becoming a nuisance.

CHAPTER

3

Next to President of the United States . . .

IT SEEMED IMPOSSIBLE: after thirty years in steel his career suddenly was falling apart. Resigning the first vice-presidency of U.S. Steel certainly did not fit the usual pattern of his life. Until now, despite occasional problems, events generally had propelled him in a single direction—upward. Indeed, in recent years Dickson had every reason to believe that some day he might become president of U.S. Steel. That such was the chief goal of his professional life can not be proved, but he once declared the office to be, next to that of president of the United States, the most important in the land. Moreover, he had always called himself an "idealist." To him that meant his aspirations—however impractical, however unrealistic, however impossible they might seem to others—in the end would be realized.

He never tried of telling how, throughout his life, his dreams repeatedly had come true. "As a boy of thirteen," he wrote, "I worshipped at the shrine of the prettiest and most beloved girl in our village." In spite of his poverty, strong objections from her family, and a rich competitor for her hand, he eventually won out and married the girl. Again, born to wealth, his father's sudden loss of both health and fortune had obliged Dickson to leave school while still a child to find work. Yet he had risen from the position of common laborer in the Homestead Mills at age fifteen to a junior partnership in Carnegie Steel by the age of thirty-three. Nine years later he was first vice-president of U.S. Steel. That he regarded as the realization of yet another "impossible" dream.[1]

Dickson's most recent "impractical" goal—hours' reform in the steel industry—had met only limited success with the enforcement of the rule against unnecessary Sunday labor. The tactics he had used to achieve even that little, however, had abruptly dampened his prospects for advancement. In early January 1911, when Corey resigned as president, Dickson was passed over in favor of James A. Farrell, a lesser official of the corporation who had not previously been regarded as a contender. Within a week Dickson notified Gary that he intended to resign. Dickson could have used the occasion to advance the labor reforms that he had championed. Certainly his position as an insider would have carried weight if he had denounced the corporation for refusing to abandon the long shift and the seven-day workweek. Unlike most critics of the industry, he could not be dismissed as an impractical "do-gooder" who knew nothing about steelmaking or the operation of U.S. Steel. Few critics knew the industry better. But Dickson instead agreed to give both Gary and himself time to reflect on the matter. More than a month later, after several meetings, Gary announced that Dickson had resigned, effective May 1.

By leaving quietly and politely, giving no reasons for quitting, and by allowing his chief opponent to announce it to the press, Dickson rendered his resignation meaningless. Perhaps he was deterred by loyalty to the industry he had served for so many years and which, in turn, had richly rewarded him. Perhaps he suspected that an outburst would appear to be sour grapes at having been passed over. He may even have nursed a hope that by resigning in a dignified manner, the corporation, in time, would mend its ways and invite him once more to its highest councils. If so, he might still—someday—head U.S. Steel. For whatever reason, he let the opportunity slip.

As Dickson explained privately, he resigned from U.S. Steel for three reasons: (1) "being a Carnegie man," he was "persona non grata" with the controlling interests of the firm, (2) he was "not in sympathy" with Judge Gary's noncompetitive commercial policy, and (3) he had "incurred the lasting enmity" of Gary by employing "drastic action" in forcing an end to unnecessary Sunday labor.[2] New York newspapers, commenting on Dickson's resignation, picked up only the first point. They noted that Dickson was the "last" Carnegie man in the upper echelons of U.S. Steel. The immediate cause of his quitting, they speculated, had been Farrell's surprising appointment as president. However, the newspapers noted, it had long been rumored in the industry that when Andrew Carnegie sold his interests to the steel trust in 1901, he had stipulated that his men were to head the firm during its first ten years of operation.[3] Whatever the truth of the matter, that decade ended early in 1911. Over the years Carnegie men in high positions at U.S. Steel had left one by one. Now, coinciding with the close of the ten-year period and the passing over of Dickson, U.S. Steel's presidency passed out of the Carnegie succession. But Dickson was right. The manner by which he forced the Sunday labor issue had antagonized Gary. So had his repeated attacks on the heart of Gary's program, the commercial policy of U.S. Steel.

* * *

It was the custom at 71 Broadway, during U.S. Steel's first decade, for executives of the firm to lunch together almost daily. "Our discussions," Dickson observed, ". . . were on every conceivable topic, seniors and juniors being on an equality." Gary was "most amiable and friendly with his juniors," and Dickson formed a "real liking for him." [4] It was at these informal meetings, however, that Dickson first clashed with Gary over the selling policies of the subsidiary companies. As with so much else, Dickson's stand on commercial policy in part related to his concern for its impact on labor.

The chairman deplored price competition and worked hard to eliminate it from the industry. At first the corporation allowed subsidiary companies to continue participating in pools just as they had prior to the formation of U.S. Steel. When corporation officials in 1904 ordered the subsidiaries to withdraw from all such arrangements because they were illegal, the same officials began periodic meetings with competitors to work out informal pricing agreements. In effect these new arrangements were "pools without penalties." Whereas traditional pools were formal agreements with trade allotments and cash penalties for violations, the new arrangements were "moral understandings" that members would not depart from agreed-upon prices without first giving notice to the other participants.[5]

Dickson and a few other officers had no faith in such gentlemen's agreements. "When business was normal these agreements were unnecessary; and when it was depressed, there were no gentlemen." [6] As long as prosperity reigned and everyone in steel was making money, Gary's policy occasioned little opposition. Whenever the market turned downward, the Carnegie-trained men at U.S. Steel became restless.

The corporation's first poor year was 1904, and Gary's noncompetitive policy drew a sharp critique from Dickson. Some of the larger subsidiary companies, the vice-president declared, were "less regarded by their competitors and by large buyers" than prior to the formation of U.S. Steel. The chief cause, Dickson believed, was the price agreements on a number of leading products. Economies of operation that came with consolidating the many firms into U.S. Steel had exceeded all expectations. By operating at only slightly over 50 percent of capacity, however, the corporation was neutralizing its advantage. Its costs were up, works that stood idle were deteriorating more rapidly than if used, and morale was low. The corporation was big,

Dickson conceded, but it would never be "big enough to ignore the laws of supply and demand." The agreements with small competitors served only to build them up until "in the aggregate" they had become "a powerful factor in their respective products."

During its first years, Dickson pointed out, U.S. Steel had absorbed competitors that became too aggressive. That tactic, thanks to the antitrust law, had about reached its limits. If the corporation could no longer absorb its rivals, it would have to fight them, unhappily "after having placed weapons in their hands." Since U.S. Steel would not be realizing much of a profit on its 1904 operations, Dickson thought it better to make that low profit "on a full output at competitive prices than by half output at artificial prices." The advantages, he pointed out, would include "a more efficient organization, lowest possible costs, and the prevention of the building up of additional competition." [7]

A sudden upturn in the market halted all discussion of the matter. For three years steel boomed. The corporation's net earnings in 1906 and 1907 would not be surpassed until the advent of the war market in 1916.[8] Maintaining prices under such conditions was no problem. In a seller's market the corporation's prices prevailed. Smaller firms could not charge more than U.S. Steel, and, since they could dispose of their product at the trust's prices, they had no incentive to cut. This ended, however, with the onset of a panic in October 1907. The market suddenly disappeared, leaving many producers and customers with large inventories of high priced steel.[9] Both prevailing economic theory and long practice in the industry dictated that prices be cut.

Gary, however, took prompt measures to preserve the prevailing price structure. Calling a meeting of U.S. Steel officials, he addressed them on the desirability of maintaining prices despite the shrinking demand. He then called on each of the officers for his view. "Being mere human beings with a natural instinct for self-preservation," Dickson commented, their remarks were "strikingly in accordance" with those of Judge Gary.

When Dickson's turn came, he dissented and advocated Carnegie's methods instead. Prices should be slashed to the point where they would bring in sufficient business "to keep intact the most valuable element in the business; namely, an experienced work force." At the same time, the sales department should be instructed to accept no long-term contracts at reduced prices. By selling for immediate delivery only, the corporation would be in a position to take quick advantage of any sudden business upturn to restore prices.

Dickson, at this first meeting on the subject, found himself standing alone. He asked Gary how he proposed to handle the problem of greatly reduced demand. The chairman replied that he would reduce output. When the meeting broke up, telegrams were dispatched to eight steel plants in Pennsylvania, Ohio, and West Virginia ordering a shutdown of operations. By December 5, U.S. Steel had blown out fifty-one of its ninety-seven blast furnaces and banked six others. The eight mills remained closed for an average of thirteen months.[10] Dickson was incensed. As he wrote years later:

> It requires no vivid imagination to predict the effects of this action on those communities. These plants had been in operation for years before the organization of the Corporation, and most of them were the mainstay of the communities in which they were located.
>
> It is true that, as compared with more modern plants, their costs were high; but if they had remained as independent plants, no doubt they would have kept some of the life-blood of wages, in circulation.
>
> Instead, the merchants were driven out of business, real estate values were depreciated; and the workmen were thrown on their own resources, having to break up their homes and seek employment elsewhere.[11]

Dickson was disturbed that in reaching so important a decision the corporation had not consulted those most vitally interested. The workmen, in fact, were not even aware that the momentous decision regarding them was being made.

> They had lost the old American status of family economic independence and sina qua non of our boasted American Democracy [Dickson noted]. By the arbitrary act of a man in a New York office, the devotee of an unworkable trade theory, their means of support were ruthlessly destroyed; and they had no redress as the strictly legal right to close the mills could not be questioned.[12]

Gary, supported by apparent near unanimity within the corporation, next carried the doctrine of price maintenance to the industry at large. As he later explained to the Stanley Committee, the problem involved the avoidance of two pitfalls. "We had no lawful right, as I understand it," he explained, "to make any agreement express or implied, directly or indirectly, with our competitors in business to maintain prices." At the same time, he believed "we had no moral or legal right to become involved in a bitter and destructive competition, such as used to follow any kind of depression in business among iron and steel manufacturers." That would produce a "war of the survival of the fittest" in which many of U.S. Steel's competitors would have been eliminated. To avoid "securing a monopoly" by driving competitors out of business and entering into an illegal agreement to maintain prices, Gary sought a middle ground. He periodically called together the leaders of the steel industry for an "interchange of information" to prevent "the demoralization which otherwise would have resulted." Once steel producers learned what their rivals were doing and how they proposed to act in the future, they saw in most instances the wisdom of not engaging in "the extreme, the unreasonable, the bitter, and the destructive competition which used to exist." [13]

In practice, Gary's methods involved more than the mere exchange of information. He hosted the first of his famous dinners for the leaders of the steel industry on November 21, 1907. There he exhorted them "like a Methodist preacher at a camp meeting." [14] Much of the success of this technique rested on telling the listeners what they wanted to hear. They must not lose their heads and give away their business by rashly dropping prices, Gary told them. Such cuts would benefit no one since, even at sharply reduced prices, few new orders could be expected during the general business collapse then in progress. "Prices should *always* be reasonable," he declared. Just as increasing demand had not been used to justify raising prices in the past, neither should falling demand now be used as a reason for cutting prices. "What we want is stability—the avoidance of violent fluctuations!" [15]

One by one the other steel manufacturers arose to endorse Gary's pronouncements. Only demoralization, harmful to all, would result from efforts "to capture an undue share of business" or to invade a "rival's territory." Declaring their faith that underlying conditions were sound, they predicted an early end to the recession. Several speakers took up Gary's argument that since they had not checked demand during the recent prosperity by raising prices overmuch, now they should not cut prices. "A slaughter of values" would not stimulate buying. Stable prices were in the best interest of producers and buyers alike so long as "they have assurance that others do not have advantages over them."

That, of course, was the crux of the matter: how to be certain that everyone would maintain the prices they all favored when accumulated surpluses invited price cutting. To handle that problem, the assemblage set up a five-man committee, headed by Gary and including Corey, president of U.S. Steel, Powell Stackhouse, president of Cambria Steel, E.C. Felton, president of Pennsylvania Steel, and Willis L. King, vice-president of Jones & Laughlin Steel Company. Anyone interested in the steel trade at any time could appeal to the committee "for advice." Through the committee, steelmen were to "cooperate for the best of the industry and conciliate any differences arising." [16]

Interviewed in England in August 1909, Gary was asked how he had so successfully overcome the "natural jealousies" of the competing firms during the previous eighteen months. "It was done by a friendly but positive effort on the part of those in charge," he responded, "to distribute the business voluntarily on a basis that was perfectly fair and just between all concerned, dependent on their capacity to produce." Enforcement lay with the committee to which the various firms could appeal "for information and advice." When a manufacturer believed that he was being unfairly treated by competitors, he complained to Gary. "I would request the company complained of to send their respresentatives to my office," Gary explained, "and we would talk it over and undertake to satisfy them they were making a mistake." Without exception, he said, the difficulties were straightened out.

Gary insisted that no agreements to maintain prices were ever drawn. Such would have been illegal, he explained. However, what legally could be and was done was "to fully inform one another of exactly what we were doing." If anyone decided to change prices, the others were notified. "The effect was that no prices were changed."

"So that you practically pooled the iron and steel business of the country," the interviewer observed.

"That is what it amounted to, although we do not call it that in our country," Gary replied. "It would not be allowed in America, as being contrary to law." [17] To Gary, an illegal pool had not been formed because it was not called a pool. A member of the Stanley Committee in 1911 asked, "You think that the Sherman antitrust law was directed at [the act of making an] agreement rather than the result of the agreement?" "I think so;" Gary replied, "I do, really." [18]

All had not been as harmonious among steelmakers between November 1907 and July 1909 as Gary's remarks suggested. The faithful had left the Judge's first dinner party fired with enthusiasm for holding firm on prices. Once back in their offices, faced with the reality of their order books, the old temptation reappeared to shave prices to make sales. When the predicted return of prosperity failed to materialize in early 1908, murmurings for price cuts increased not only among competitors of U.S. Steel, but within the corporation itself.

Gary, well aware of the nature of the sin besetting the industry, reassembled the steelmakers in May. Once more he proclaimed the gospel of harmony, cooperation, and stability, and urged backsliders to resist falling into error. And again he was successful. "The impression had been gaining ground," *Iron Age* commented afterward, "that a disposition to reduce prices had been increasing in the ranks of the manufacturers." Following the meeting, however, Gary announced that "the opinion was expressed by each one present that the prices of steel are reasonable and should not be reduced; that reduced prices would not increase purchases; and that most of their customers do not expect or desire any changes." That night Gary once more entertained his guests at the Waldorf-Astoria.[19] "Gary and I," Perkins reported to Morgan next day, "really had more trouble within our own ranks than with outsiders." [20]

Whether or not Gary's scheme would ultimately succeed rested on fear, fear that in any real competitive test the corporation could overwhelm its rivals. "It is clear," Gary told a *Chicago Tribune* reporter, "that the United States Steel Corporation, with its extensive resources, could use its great strength, like other corporations, to crush competition." That it did not, stemmed less from good will and friendship than from knowledge that such a policy would only bring U.S. Steel under attack for violating the antitrust laws.[21] Still, to be convincing, the corporation could not allow smaller firms to cut prices with impunity. When, late in 1908, the small independents began to undersell the larger companies, Gary's program showed signs of crumbling.

Dickson attacked from inside. In a letter to Corey on February 16, 1909, the vice-president admitted that Gary's policy, "generally speaking," had been successful over a considerable period of time. However, in spite of the corporation's relatively high earnings, it had suffered indirect losses that were difficult to measure: the build up of competitors, demoralization of the sales force, and the enforced idleness of employees due to prolonged shutdowns of some plants and reduced crews at others. Support for Gary's policy came chiefly from the larger steel companies, Dickson observed. Small mills had cut prices and taken over "an undue share of the business offering." So common had price cutting become in nearly every line, he continued, that U.S. Steel subsidiaries found themselves practically isolated in their attempt to maintain official prices.

The corporation could no longer ignore the combined impact of the small companies on market prices, Dickson argued. It must "either advance to a higher degree of cooperation or return to competitive conditions." The first was possible only by resorting to illegal methods, "which, of course, are not to be considered." The alternative was for the corporation to use "quality of material, fair, and if necessary, low prices, diversity of product, and efficiency of service" to retain its share of business "in open and friendly competition with the world." [22]

Two days later, following a series of meetings with the leaders of U.S. Steel and the larger steel manufacturers, Gary surrendered. "To protect their customers," the major firms for the present would "sell at such modified prices as may be necessary . . . to retain their fair share of the business." [23] With good reason, Gary's later statement that U.S. Steel would seek orders "at any price that can be obtained," struck concern among the independents.[24] "Nearly everything placed in the last week was taken by us," the sales manager of Carnegie Steel reported to 71 Broadway in mid-March. "The other mills seem to think the pace we are setting is a little too fast for them." According to *Iron Age,* the question—quiet since the formation of the U.S. Steel in 1901—of whether even large independents could compete with the corporation in a lengthy test of strength was "being canvassed again." [25]

In fact, Gary had not abandoned his goals of harmony and stability, nor was he precipitating a ruthless price war in steel. Rather, he was using open pricing to teach a few lessons to his opponents both in and out of the corporation. The rejoicing of his Carnegie-trained colleagues at U.S. Steel was shortlived. The sudden rush of eager salesmen into the open market brought back no flood of orders such as Dickson and others had predicted. The move came far too late in the business cycle for that. Steel had bottomed out much earlier. The corporation's average daily orders in December 1907 had been but 8,322 tons. Orders gradually increased to 18,349 tons per day by June 1908, and to 29,685 tons per day by December. The opening of prices in February 1909 increased orders, but only to a little more than 30,000 tons per day. The gains, in other words, simply continued an existing trend rather than marked a revolutionary improvement.[26]

Even though orders had not vindicated their position on prices, Corey and Dickson proceeded to attack Gary's policy of steady wage rates. So long as prices were unchanged, Gary insisted, fairness to workmen required that wage rates remain steady too. During the recession, wage rates at U.S. Steel did not change in spite of a slight dip in the industry at large. Once fixed prices were abandoned, however, several steel companies announced wage cuts of approximately ten percent.[27] Corey, Dickson, and their allies insisted that the corporation must follow the trend. Most of the steel producers thought of the cuts solely in terms of reducing costs. Dickson, as already noted, hoped to stimulate orders by lowering prices and then employing most if not all of corporation's predepression labor force. Gary's repeated boast that wages had not fallen at U.S. Steel seemed a mockery to Dickson, considering that the corporation had closed down entirely many of its plants. The average labor force, in fact, had been reduced from 210,000 in 1907 to 165,000 in 1908—a 21 percent cut. Thanks to layoffs and part-time operations, the average annual take-home pay of employees who kept their jobs fell by nearly 5 percent—from $765 in 1907 to $729 in 1908.[28] Having men at work, even at reduced rates, Dickson reasoned, was to be preferred to having them idle, however high the wage rates of those with jobs.

In early March 1909, Corey appeared before the finance committee to plead the case for lowering wages. Among the bankers who dominated the committee, however, Gary's arguments prevailed.[29] Sentiment was divided, nonetheless, and in early April the press reported "on the best authority" that subsidiaries of U.S. Steel would announce wage cuts on April 15 to take effect on May 1. "Well, we see you had your own way after all the talk," a colleague noted on the clipping that he sent to Dickson. "I argued at the meeting that lower wages were better than no wages. The alternative proposed was to shut down some mills," Dickson recorded.[30] But U.S. Steel did not cut wages. Gary appealed the issue to Morgan, and the banker suggested that any changes be postponed to await further business developments. When orders picked up in May, the finance committee voted to keep wages steady.[31]

As soon as orders increased, Corey demanded that prices be raised. Again Gary refused. The independents still had to be taught the folly of trying to cut into the corporation's share of the market by shaving prices. And so, although competitive pricing increased the flow of orders to most steel companies, only U.S. Steel could make a profit at the new price levels.[32] Independents, moreover, could not hope to raise their prices and sell steel so long as U.S. Steel kept its prices down. How long they would operate without profits was wholly in the hands of Gary.

The Judge, of course, had no intention of allowing the rival firms to fail and thus stir up the proponents of trust-busting. During the summer of 1909 he once more began preaching the doctrine of harmony and called for reasonable, stable prices. Gradually the corporation (followed quickly by the independents) edged prices upward. By fall, Gary was the hero of the industry. Thanks to his policies, most companies had disposed of their inventories without loss between November 1907 and February 1909 while prices were fixed. U.S. Steel's impact on the market during the period of open pricing, while strong enough to frighten the independents, had produced no marked or lasting change in any company's overall share of the market. By September most steel prices had returned to January levels or higher.[33] All had weathered the panic and were in good shape to take advantage of returning prosperity.

In October the industry paid homage to Gary at a banquet at the

Waldorf. Over one hundred steelmakers were there. Corey, for one, did not attend, but Dickson and most of Gary's other opponents were present when the leading steel manufacturers arose to praise Gary for what he had done. He had taught them, one observed, that "what is good for my competitors is good for me." "I can imagine," he continued, "an old timer, who had gone down to ruin in the old days, crying out, 'Oh to be in business now that Gary is there.' " [34] Willis King of Jones & Laughlin declared: "I think we like him best for showing that a man may be in business, a large business, and still be a gentleman." [35]

The climax came when Schwab, president of Bethlehem Steel and a long-time champion of Carnegie methods, presented Gary with a gold loving cup. "I am thankful for this opportunity of saying one thing, Judge Gary," he began. "You and I were associated in business for some years. We had many differences, and I am glad of this public opportunity to say that with my bounding enthusiasm and optimism I was wrong in most instances—indeed, in all instances, and you were right. The broad principles that you brought into this business were new to all of us who had been trained in a somewhat different school. Their effect was marvelous, their success unquestioned." [36]

Gary responded briefly, deprecating the days of ruinous competition and hailing the new era of cooperation. Gone was the time when "a competitor was treated as the common enemy." Under the new dispensation "you can witness the success and prosperity of your neighbor without the slightest feeling of envy or discomfort. You believe in competition, but not hostility; in rivalry, but not antagonism; in progress and success for all, but not in punishment or destruction of any." [37]

Although Gary stressed the positive side of working together, one sentence hinted at a darker sentiment. "Bitter, wrathful and destructive competition in business," was no longer permissible "*unless it becomes necessary in self-defense.*" [38] At a later time, in the privacy of the luncheon meetings at 71 Broadway, the chairman indicated that he was more than willing to punish those who refused to support his policy of continuous cooperation within the industry. When a company, such as Inland Steel, refused to go along, Gary said in January 1910, it should be disciplined by those who were cooperating. In an even less charitable mood in November 1910, Gary declared that "certain mills were not entitled to live" and recommended that those willing to cooperate should band together to "train their guns" on the offenders. In both January and November, when Dickson objected that such actions would constitute conspiracy, Gary answered that the measures could be justified before federal authorities if necessary. At the November luncheon, he claimed that President Taft and Attorney General Wickersham had given him assurances that such reprisals would be proper and just.[39]

* * *

Gary had won a complete triumph at the banquet in October 1909, and from that moment on Corey's and Dickson's challenges to his authority and policies were futile attempts to reverse a running tide. A month after the banquet a situation arose over the power of a subsidiary president that not only led to a power confrontation but laid the groundwork for both Corey's and Dickson's eventual resignations. In November 1909, Gary invited John C. Greenway, general superintendent of the corporation's Oliver Iron Mining Company in Minnesota, to come to New York to discuss the subsidiary's affairs with himself and Perkins. Since the meetings were to be confidential, Greenway told his superior, President William J. Olcott, that he had personal business to transact in New York. Upon his return, Greenway said nothing about the discussions, and Olcott learned of them only by accident. Summoning Greenway before him, Olcott charged him with disloyalty, asked for his resignation, and when refused, fired him.

Greenway appealed to Gary. The chairman discussed the matter first with Perkins and then with Perkins, Olcott, and Corey at a lengthy meeting on December 29. Gary and Perkins, concerned for the authority of the corporation's officers over subsidiary officials, insisted that Greenway be reinstated. Corey stood firmly behind Olcott, trying to protect his own and the subsidiary presidents' waning autonomy and power. Reinstatement of Greenway would undermine internal discipline in the subsidiaries and reduce the presidents to virtual ciphers. No decision was reached at the heated session, but afterward Perkins immediately conferred with J. P. Morgan, Jr., and finally with the great financier himself. Perkins, with Morgan's support, telephoned Corey the next day to learn if he had changed his mind. Corey had not, and warned that if Greenway were reinstated, Olcott and other officials of the mining subsidiary would resign. Perkins assured Corey that the resignations would be accepted.[40]

On December 31, Gary sent Corey a note instructing him to "recommend" to Olcott that Greenway be restored to his position. Corey refused and asked that the matter, which was of "most vital importance" to the corporation, be brought before the entire finance committee for resolution. Gary preferred settling the affair himself. "I certainly think you are wrong in your conclusion and in your position," he wrote. "I believe if you will read the bylaws carefully you will agree." He suggested that Corey meet with him and in the meantime withdraw his letter of refusal and "promptly comply" with his "request" regarding Greenway.[41]

Corey refused to be intimidated and a meeting of the finance committee was set for January 4. Gary and Corey set about lining up supporters. What had begun as a minor incident now escalated into a full-blown confrontation between the steel producers and the lawyer-bankers. Assisted by Dickson, Corey campaigned for support among his natural allies, the subsidiary presidents. At Corey's urging, these officials began arriving in New York immediately after New Year's Day. Meanwhile, Dickson drew up a petition addressed to Corey. "Next to its ownership of raw materials," the document declared, "the greatest asset of the Corporation is its organization, and any condition which imperils the efficiency of the organization through the weakening of the proper authority of any officer, should receive careful consideration." Reciting a brief account of the Greenway incident, the petition protested Gary's actions as a "radical change" in the administration of the corporation's affairs. To prevent a repetition, to assure future harmony, and to safeguard the welfare of the corporation, the petition recommended that the respective duties and authority of the chairman and of the president of U.S. Steel be "clearly defined." Affixing his name at the top, Dickson recruited the signatures of fifteen other officers and directors, including eleven of the twenty subsidiary presidents. Seven of the sixteen signers, including Dickson, were Carnegie veterans.[42]

Dickson also seems to have prepared recommendations for Corey's use at the meeting of the finance committee. President Olcott should be allowed to present his case in person, Dickson suggested, and the various subsidiary presidents then in New York should be called upon to give their views. The key point, however, was to emphasize that the present organization of the corporation was "a superb machine unexampled in its efficiency in the history of the world." The ability of the subsidiary presidents to construct and operate the "vast enterprises" over which they presided was largely due to "loyal organizations" that they had built up over the years and to the "mutual confidence and support which each president received from the officers of the Corporation, from his fellow presidents, and above all, from his subordinates."[43]

Corey, meanwhile, tried to recruit support among members of the finance committee. In this contest with Gary and Perkins, who were backed by Morgan, Corey had no chance, losing even Henry Clay Frick—the one-time Carnegie partner now sitting on U.S. Steel's board of directors—who by training and instinct inclined to the steel producer's point of view. At the three-hour session on January 4, Olcott spoke of the demoralizing effects of Gary's methods. The other subsidiary presidents were not invited to attend, much less to speak. Alternately meeting as a group and then breaking up into subgroups, the committee debated the issue. When Frick finally took Corey aside and told him that support for Gary was unanimous, Corey agreed to execute any order the committee handed down.

Perkins, wanting the question of authority settled once and for all, was ready to call for Corey's resignation. Decision on that issue was deferred. After a unanimous vote to order Greenway's reinstatement, Perkins asked the chairman if there were anything more that he would like done. "Yes," Gary replied, "I should like a resolution more definitely fixing my status in this Corporation." A resolution was drawn and subsequently adopted by the board of directors on March 1, declaring the chairman to be "the chief executive officer of the corporation, and, subject to the Board of Directors and the Finance Committee, . . . in general charge of the affairs of the Corporation." [44]

The full significance of the finance committee's decision did not escape Dickson. On January 5, 1910, he prepared his resignation and a letter to Corey. "While I expect you to use your discretion as to the time of its acceptance, I desire to make it clear to you that the document means just what it says and that there are no strings to it. . . . I have very grave doubts as to the permanency of the settlement reached in the unfortunate controversy which has arisen and fear that our position may be made so untenable as to force us to decide between the sacrifice of our position or our self-respect." Dickson delayed sending in his resignation until February 1 at the request of David G. Kerr, second vice-president of U.S. Steel, a signer of the petition, a Carnegie associate, and close friend. As matters turned out, the rebels were not asked to resign and none did—at least for the time being. Corey was to hold on for another full year, quitting on January 31, 1911. As he left he returned Dickson's unused resignation to him.[45] With the presidency vacant, Dickson may have again thought of himself as a possible contender, and so submitted no new resignation. That would come in March 1911.

* * *

Early in 1910, the informal Gary dinners were expanded and institutionalized into an organization called the American Iron & Steel Institute. George G. McMurtry, chairman of the board of American Steel & Wire (a subsidiary of U.S. Steel), belonged to the British Iron & Steel Institute and regularly touted the benefits of membership in the

British organization to the men lunching together at 71 Broadway. When he raised the possibility of forming an American counterpart to the British organization, the suggestion was acted on by Gary, who liked the idea and called an organizational meeting for May 27 after consultation with other steel executives. Both for that meeting and the first formal session of the American institute in October, Gary asked Dickson to be a principal speaker.[46] If the invitation was meant to give Dickson an opportunity to prove that he had learned from recent events, the vice-president failed on both occasions. He used the meetings as platforms to promote labor reforms in the industry at large.

At the May session, Dickson observed that the British institute limited the topics brought before it to technical and scientific questions and that special care was taken to avoid any discussion of commercial matters. Dickson urged that the American institute permit both commercial and sociological questions. In the latter category, he suggested, a matter most deserving of "immediate and earnest thought" was employer-employee relations. U.S. Steel had made important advances in recent months, he noted, having reduced the seven-day workweek to a minimum, established a system of accident and accidental death relief, and set up pensions for employees too old to work. On the matter of the seven-day week, he admitted, the corporation had not yet solved the problem for blast furnace operations. He invited the cooperation of the other companies in devising a workable plan for the industry as a whole.

Prompt action by management in the area of employer-employee relations was imperative, Dickson argued. He pointed to the recent introduction and passage of employer liability laws in several leading industrial states and observed that the federal House of Representatives had just added to a bill that authorized construction of two battleships a proviso that the materials used had to be the product of eight-hour employment. Should the steel industry, he asked, attempt to pass through the "troubled waters with sufficient headway to insure steerage way," or would it "be satisfied to drift, taking [its] chances with disaster on the sunken rocks of radical and ill-advised legislation?" True conservatism, Dickson told his audience, called for an adjustment of their operating methods to changing conditions. The present system, with its twelve-hour day and seven-day workweek, was "a reproach to our great industry and should not in this enlightened age be longer tolerated." He urged the group to name a committee to consider the problem and to report a workable plan to the industry "whereby no individual shall be on duty for more than six consecutive days." Such a course was not only right, it was also in the spirit of the age. "If we do not do it voluntarily," he warned, "we will in the near future probably be compelled to do it by the passage of legislation . . . which may be so radical as to create a serious situation for the entire Iron and Steel industry." [47]

The assembled steelmasters had no way of knowing that Dickson was not speaking for Gary and U.S. Steel. When called upon for comments, Schwab voiced approval of the proposals but warned of the costs involved and the difficulty of securing sufficient workmen to cover three eight-hour shifts per day. Others, including George W. Perkins, believed that the institute would be able to do much toward solving the questions raised by Dickson. Finally, James Clarke, of Youngstown Sheet & Tube, moved that the directors of the institute appoint a committee of five to investigate seven-day labor as Dickson proposed.

Gary, who spoke last, clearly was unhappy at the course the discussion had taken. He, for one, he said, did not favor taking up sociological questions such as the one under consideration—especially out of any sense that public sentiment compelled them to do so. To him, fidelity to the interests they represented was important. "The highest type of an honest man is not the one who leans over backward in his effort to be fair, but the one who is not afraid to decide in favor of his own friend if that decision would be right." The American Iron & Steel Institute must never allow itself to be pushed into a stand without regard for whether the public sentiment in question was right or wrong.[48] Despite passage of the Clarke resolution and praise of the proposal in the industry's leading journals, Gary left for vacation in Europe without acting. Only after his return and at Dickson's "earnest solicitation," did he name the committee.[49] This did not prevent him from boasting at the Stanley hearings in 1911 that, although the institute took up technical matters, it also pursued "many other questions, ethical questions, sociological questions and questions that nowadays . . . thinking people believe to be as important as the mere question of making a little money." High on the list was the "welfare of our men." That subject, he noted, had been "one of the principal discussions" at the first two sessions of the institute.[50]

At the first formal session of the institute in October, Gary indicated his view in his opening address as to the organization's real purpose. "Primarily the Institute was organized, and should be conducted, for the benefit of its members," he declared. "It should result in decided pecuniary advantage to all." There were many pressing topics to be considered. "One of the most important business matters at this time, and perhaps at all times," he said, was "the maintenance of stable conditions in the iron and steel industry; and to this is attached the question of prices." After discussing the benefits of steady, fair prices, both to the industry and to its customers, Gary outlined how the goal was to be achieved: "Frank and friendly intercourse; full disclosure of his business by each to the others; recognition by all of the rights of each; a disposition to assist and benefit each other so far as practicable and proper; [and] conduct founded on the belief that healthy competition is wise and better than destructive competition." The balance of Gary's remarks and those of the next speaker, James A. Farrell, president of the corporation's Products Export Company, were directed to the foreign steelmakers who were guests that day. The cooperation that had worked so well in the United States during the recent panic could and should, they believed, be extented to the industry worldwide.[51]

When his turn came, Dickson once more focussed attention on the question of employee relations. No one with experience in the industry would minimize the importance of keeping workmen "contented with their lot," he declared. Only recently had the subject expanded beyond wage rates to the larger question of the "social responsibility" of employers. Dickson sketched out U.S. Steel's recently initiated safety and accident relief programs and its pension scheme for employees—all of which he had helped to develop. In the course of his address, Dickson uttered some surprisingly radical ideas for a meeting of so eminently conservative a group. He complained, for example, of the "law's delay" in settling accident claim cases. Legal technicalities, retrials, and appeals to higher courts all worked to "the immense disadvantage" of men of limited means. "To our shame," he declared, "we must admit that under such conditions the poor man and the rich man or corporation, are not 'equal before the law.' " Commenting on the new employer liability acts, he stated his belief that "compensation to injured workmen" was "a legitimate charge against the cost of manufacture," and that an injured workman and his dependents should "receive compensation, not as an act of grace on the part of his employer but as a right." The ultimate solution, he predicted, would come in the form of a tax on all employers, based on the number of employees and "the hazards naturally inherent in each industry."

Once more Dickson brought up long hours, the seven-day workweek and what had to be done to end their incidence in the industry. The problems were not easy, he conceded, but they were less difficult than many that the industry had already solved. Having been so successful in developing "mechanical appliances and processes" and in administering immense enterprises, he had no doubt that the leaders of the industry could also solve the social problems facing them "when they are confronted with the absolute necessity for their solution." [52]

Gary had made no effort to control what Dickson might say. He had taken the precaution, however, of asking three men to comment on Dickson's paper. Charles Kirchhoff, the respected former editor of *Iron Age*, praised Dickson's stand. The other two, Edgar S. Cook, president of Warwick Iron & Steel, and Edward Bailey, president of Central Iron & Steel, had been chosen to attack his views, Dickson suspected.[53] Cook did. Considering the class of laborers employed at blast furnaces, he declared, it was a waste of sympathy to be concerned about their working twelve-hour shifts, seven days a week. Work was not continuous, he said. Moreover, what would the men be doing if they were not working? "They have just enough of the beginnings of education to get together and make themselves unhappy and discontented," he said, "but with no resources to occupy their time profitably." He was certain that "eight hours would be to their unhappiness, to our discomfort, and to the benefit of no one."

> The blast furnaces [he continued] may be likened to a preparatory school, fitting boys for a college course, or business career. A wise master knows that the boys must be kept busily employed, either in recitation room, study, or play. All must be laid out according to rule, so that every hour of the 24, including the hours of sleep, has its particular duty. Lacking such an organization, the pent-up energies of the boys would concoct all sorts of mischievous schemes, working injury to themselves, and destruction to the school. Many men never get beyond the school age, and it is necessary for their protection and advancement, that they should be surrounded by an organization that will guide them along the lines best suited for their capacities.

As for the seven-day workweek, Cook continued, it was no problem. The men "generally manage to take a day off whether with or without the consent of the foreman." Only the twenty-four hour turn was a hardship, he said, and for that he saw no remedy.[54]

Bailey, who said that he had "been asked to criticize" Dickson's remarks, conceded that he had "nothing but the heartiest approval of everything he has said, from the humanitarian as well as the economical point of view." Dickson once more had scored, in his listener's consciences if not in their deeds. It was to be his last victory at U.S. Steel.[55]

* * *

The title, President of the United States Steel Corporation, conveyed an illusion of power not wholly in accord with fact. In setting up the corporation in 1901, J.P. Morgan had conferred the title on

Schwab, but had also restrained the steelmaker's authority by setting up an executive committee of the board of directors, chaired by Judge Gary, that effectively held Schwab in check. Failing in his efforts to free himself from control by the committee during the strike of 1901, Schwab found his powers eroding. Criticism of his personal life and extra-corporate dealings further weakened the office by the time Schwab turned it over to Corey in 1903. As matters turned out, scandal in Corey's life would continue the process to the point where the board of directors briefly considered abolishing the office altogether following Corey's resignation.

Shortly after moving from Pittsburgh to New York City in 1903, Corey's wife sued for divorce. Corey had deserted her for a chorus girl after twenty years of marriage. The divorce proceedings, Corey's second marriage, and stories of his lavish spending on the new Mrs. Corey repeatedly drew front page attention in the New York press between 1905 and 1908.[56] His standing within the corporation could not have been helped by recurrent rumors that he was about to resign or be replaced. One reporter, assuming that Corey was about to leave and that the succession would remain in the Carnegie line, mentioned two likely contenders: Dinkey, who had succeeded Corey as president of Carnegie Steel, and Dickson, "one of the most thorough steel manufacturers in the country." [57]

Because Gary stood firm in his support (the chairman may have preferred a vulnerable president), Corey weathered the divorce crisis. Meanwhile, Dickson advanced a notch closer to the top, thanks to the marital difficulties of another high-ranking Carnegie man at U.S. Steel. First Vice-President James Gayley and his wife separated in November 1907 after twenty-four years of marriage. Mrs. Gayley took up residence in Rome where one of the Gayley daughters had recently married an Italian count. With Corey's affairs still making headlines, Gayley elected to resign quietly. Dickson succeeded him on January 1, 1909.[58]

Corey's usefulness at U.S. Steel came to an end with his defeat in the Olcott-Greenway imbroglio. As early as April 1910, Morgan concluded that he must go. For one reason and another, Corey remained on till the end of the year. In all likelihood, Gary had already begun to take measure of potential new presidents. He may have dropped Dickson from consideration long before. If not, the vice-president no doubt clinched the matter by his use of the American Iron & Steel Institute meetings as a forum for his labor views and his declaration to Gary at lunch one day in November, 1910, that he "did not now nor ever did believe in the so-called cooperative movement." [59]

Corey submitted his formal resignation on January 3, 1911. The finance committee, in reporting the event, noted that it would not choose a successor "in the immediate future, if at all." Frick, among others, questioned the need for a president.[60] One week later, however, Gary announced that James A. Farrell, president of the U.S. Steel Products Export Company, had been selected for the presidency of the corporation. This sudden reversal reportedly was due to the flood of detail work that otherwise would have fallen to Gary. According to a corporation official, Farrell was "admirably suited for such work." He was "a good work horse," and could be depended on "to follow the policies of the Corporation." [61]

The announcement took Wall Street and the industry by surprise. Most newspaper accounts had picked Dinkey, Dickson, or Buffington of Illinois Steel as the logical successor to Corey.[62] "There was hardly a man to be found" who had ever heard of Farrell. About all that reporters could learn was that he lived in Brooklyn, "which in itself was enough to confirm Wall Street in its opinion that the controlling interests in the Steel Corporation had determined to make sure that the personal affairs of its high officers should cease to be intermittently the subject of newspaper headlines." [63]

Dickson probably first learned of Farrell's appointment at the January 10 annual dinner of U.S. Steel officials. The next evening the new president was the center of attention at a dinner for the American Iron & Steel Institute. There Dickson listened as Farrell declared:

> I understand the policy of the corporation to be to co-operate with its competitors in the effort to maintain fair prices, and the stability of business conditions, by every means permissible under the laws of the country and not antagonistic to the public conscience. I claim to be the good soldier. Knowing the policy of the corporation, I recognize the obligation on my part to carry out to the fullest extent its parts as defined and interpreted by Judge Gary, the chief executive officer; and I may be expected to follow that course for that reason.[64]

Gary clearly had selected his kind of president for U.S. Steel–a man who would carry out orders without question—a "good soldier." Dickson would have used such an occasion to announce some bold new scheme, not to pledge fealty to the chairman. And that, of course, was one of the reasons why Farrell was speaking, and not Dickson.

Five days later, Dickson met with Gary and offered to save him and Farrell "any embarrassment" by resigning. Gary reviewed the considerations of the finance committee in selecting Farrell and told

Dickson that it would be a mistake for him to quit. Dickson replied that, having studied Farrell's methods during the past decade, he doubted that he would be able to maintain "pleasant relations" with the new president. Gary said that he wished to discuss the matter with Farrell before deciding.[65]

At a second meeting on February 6, Dickson told Gary that he had concluded that the new administration would be unable "to utilize my services to the same extent which Mr. Corey had." Under the circumstances he preferred not to remain "unless I felt I was rendering value received." Since the changed conditions were not his fault, Dickson suggested that the corporation should not penalize him by retaining certain shares of stock belonging to him. Gary, responding "in the most kindly manner," assured him there would be no question of his receiving the stock. The chairman, however, wished to discuss Dickson's furture with others, particularly Frick and Morgan, Jr. On February 9 he telegraphed Dickson, urging him to delay his decision.[66] Dickson did not. On March 17 Gary made the resignation public.

In commenting on Dickson's departure from U.S. Steel, the New York press predicted a busy future for him. One set of stories suggested that the various Carnegie men driven from U.S. Steel—Schwab, Corey, Dickson, and Dinkey (his resignation was predicted as coming momentarily)—were about to join forces. Backed by Andrew Carnegie's money, they were going to consolidate Bethlehem Steel with a number of other independent firms into a new combine second only to U.S. Steel in size.[67] Another prediction was that Dickson, "a man of wealth," would devote himself to public service, perhaps even enter New Jersey politics.[68] Neither prediction came true. No new combine was formed and Dickson never sought public office. Instead, he retired with his family to Highland Croft, his New Hampshire farm, for the summer. As matters turned out, Dickson's retirement at the age of forty-six lasted somewhat longer than he expected. His "impossible dream" of someday heading U.S. Steel had to be deferred—indefinitely.

CHAPTER

4

The Battle Continues

This was what I had prayed for: a small piece of land
With a garden, a fresh-flowing spring at hand
Near the house, and above and behind a small forest stand.
—Horace

SO WROTE HORACE of his farm in the Sabine Hills and so Dickson, who admired the ancient poet's work, spoke of his "Sabine Retreat" in New Hampshire. Retirement to quiet rustic living, Dickson told himself, was what he really craved. He celebrated his "removal from city streets to mountain retreat" [1] that first summer by abandoning himself completely to the life of a gentleman farmer. With characteristic zest he systematically marshaled his children, hired hands, and even visitors, and plunged into the work at hand. He put in a lawn and flower beds, planted trees, and cared for a garden. He laid out paths and a road across his property. He built a rough bridge and a summer house with thatched roof. He improved the pond for swimming and constructed a tennis court. As the summer passed, the whole family joined in to make hay in July and to gather apples for cider in September. The Dicksons always interspersed their work with pleasure: swimming parties, picnics, auto tours to nearby scenic spots, trips to the village for church, or a moving picture show. They went to the circus in July and to the county fair in September. And there was a steady flow of visitors: friends, relatives, and school chums of the children.

When autumn brought an end to the idyllic summer, Dickson and his family left New Hampshire, the older children returning to their school while Dickson, his wife, and the younger children returned to Montclair. The winter months were devoted to cultural affairs: the opera, concerts of the New York Philharmonic, and recitals by prominent violinists, pianists, and singers. In December Dickson attended the annual meetings of two of his favorite organizations, the Carnegie Veterans' Association and the Pennsylvania Society. He also enjoyed frequent luncheon meetings with friends at the Lawyer's Club in New York.[2]

Conditioned to hard, steady work from the age of eleven, Dickson found his relatively inactive life at forty-six all but unbearable. A summer at his New Hampshire farm could be delightful, but far more was needed to give his life purpose and meaning. "My retirement from the Steel Corporation," he later wrote, "was followed by four years of mental restlessness and vain attempts to find satisfying employment of my time." [3] In July 1911 he was offered the presidency of the Lehigh By-Product Coke Company, which he declined. In October 1912 the International Pump Company's headship fell vacant, but when Dickson asked for a salary of $50,000 with an option on 25,000 shares of the company's stock, negotiations fell through.[4]

How anxious Dickson was at first to secure new employment is not clear. In the back of his mind he seems to have clung to the hope the U.S. Steel might invite him to return. He was careful for the next several years to conduct himself as a loyal opponent of present policy rather than as an enemy of the corporation. Even as he was leaving the vice-presidency in May 1911, he refused to join in attacks on the firm. About that time the Stanley Committee of the United States House of Representatives was beginning its lengthy investigation of the Steel Trust to determine whether its formation and subsequent behavior violated the Sherman Antitrust Law. Dickson, his resignation only a few weeks old, was subpoenaed to appear. The committee, he believed, expected him to testify against the corporation. Resenting the implication that he might be disloyal, he sent word that he would obey the summons but if asked about price fixing at the Gary dinners would state that on two occasions, when Gary and Corey were out of town, he personally had presided at the meetings and on both occasions the attorney general of the United States (then an attorney for the corporation) had sat at his right hand "to advise so that no illegal action would be taken." [5] Although Dickson was called to Washington at the end of May with Gary and other corporation officials, he was not called upon to testify. His relations with his former colleagues were friendly; following Gary's testimony on June 1, Dickson accompanied steel officials to a baseball game. After two days, with the committee's permission, Dickson left for home and was not called upon to return.[6]

A few months later, when the Justice Department launched an antitrust suit against U.S. Steel, one of the attorneys for the government asked Dickson if he might call on him "to take up certain matters in connection with the Government's case." If the purpose was to serve a subpeona, Dickson replied, he could be reached at his office in New York or at his residence in Montclair on a specified date. Otherwise, he preferred not to discuss the matter.[7]

At about the same time, Dickson prepared a memorandum, possibly for use in the event that he should be called upon to testify. "I had intense pride in the United States Steel Corporation," he wrote, "second only to the pride which I had in the old Carnegie Steel Company." He admitted to disagreeing with the corporation's "so-called co-operative policy" of fixing steel prices but observed that that policy had failed "of its own weight" and doubted that any serious attempt would ever be made to revive it. He also disapproved of the corporation buying up ore lands simply to keep them out of the hands of competitors. To prevent big businesses in general from hoarding natural resources, Dickson subscribed to the principle of Henry George's Single Tax. Nonetheless, he said, U.S. Steel "in its treatment of stockholders, employees, customers and competitors, deserves well at the hands of the American people." Any action resulting in the disintegration of the corporation, he declared, "would be a grave public calamity." [8] Dickson was not called on to testify. His loyalty to U.S. Steel remained unsullied.

* * *

Beginning with his last year at the corporation, and on through his protracted retirement, Dickson played a major role in the enactment and administration of one of the nation's earliest employer liability laws. In April 1910, just four months after U.S. Steel adopted its voluntary accident relief plan, Governor J. Franklin Fort of New Jersey invited Dickson to serve as president of a state commission to study the problem and to frame an appropriate employer liability law. Dickson regarded the project as "one of the really worthwhile activities" of his life.

The New Jersey commission held extensive hearings throughout the state. To Dickson's disgust, most employers "assumed a hostile attitude and practically boycotted the sessions." When the commission submitted its report and a draft bill, the employers opened a campaign against the proposed legislation. At legislative hearings on the measure, a young attorney hired by the New Jersey Manufacturer's Association delivered a "spread eagle, American flag" speech exposing the "dangerous features of the bill." Noting that the lawyer's presentation contained gross errors of detail, Dickson asked him pointblank whether he had actually read the bill that he opposed. The young man sat down without reply and took no further part in the hearings. Dickson in turn denounced the employers for their "unreasoning attitude of blind opposition" and stressed the conservative features of the measure.[9]

The bill subsequently passed both houses of the legislature with a single dissenting vote and became law when signed by Governor

Woodrow Wilson on April 4, 1911. The new law was one of the first of its kind in the nation. It set aside the classic defenses of employers against liability claims—the "fellow servant" and "assumption of risk" doctrines—and modified the "contributory negligence" doctrine by requiring employers to prove willfull and deliberate negligence on the part of injured workmen. To protect themselves against excessive claims, many employers subscribed for liability insurance with private underwriters. This was not mandatory under the New Jersey law, however.[10]

From the beginning, Dickson saw the voluntary insurance feature as the chief shortcoming of the New Jersey law. In an address before the American Academy of Political and Social Sciences in April 1911, he noted that small employers had opposed the New Jersey statute because they feared it would impose heavy financial burdens on them. The remedy at hand—liability insurance—unfortunately was also very costly. One answer, Dickson suggested was more mutual accident insurance companies. Besides providing protection to employers, such firms—like fire insurance companies before them—would undoubtedly seek to cut costs by working to reduce or eliminate accidents and that, of course, would benefit everybody. But neither the New Jersey law nor private liability insurance companies offered a permanent solution to the problem, Dickson declared. In the near future it would be "repugnant to an aroused and enlightened social conscience that the insurance of injured workmen should be a source of profit to any one." Eventually, a system of state insurance, "compulsory both on the employer and employee" would resolve the matter, he predicted.[11]

In 1914, Dickson, still a member of the commission, wrote Governor James F. Fielder with regard to changing the New Jersey law. Although the measure at the time of its adoption had been "the most advanced legislation on the subject in the country," other states had subsequently enacted stronger laws. The greatest weakness of the New Jersey law, Dickson asserted, was that payments to injured persons continued only so long as their employers remained solvent. What was needed, he insisted, was compulsory insurance.[12]

Actually, voluntary insurance was only one of several weaknesses that marred the New Jersey and other state systems set up prior to World War I. Most of the plans provided for a waiting period before benefits began, and cash payments to the injured were low—even total disability payments in New Jersey were not allowed to exceed 50 percent of ordinary wages. Although promoted as a way to escape lengthy litigation, the New Jersey and several other systems were initially administered through the regular court system.[13] Most of these defects in employer liability laws continued into the New Deal period and beyond. Dickson's active interest in the subject came to an end in May 1916 when the commission he headed dissolved and turned administration of the law over to the Bureau of Workmen's Compensation—an agency within the New Jersey Department of Labor.

* * *

Early in his retirement, Dickson strengthened ties with reformers such as Paul U. Kellogg and returned to the struggle for hours reform that had undermined his position at U.S. Steel. When Kellogg first learned of Dickson's resignation, he telegraphed, "If my friendship is worth anything at this juncture, it is yours," and invited Dickson to call on him.[14] Writing to James Bertram, Andrew Carnegie's secretary, Kellogg noted Dickson's resignation. "Mr. Dickson is a Carnegie man," he observed. "He has done more than any other ten men in fighting for the reduction of unnecessary seven-day work in the steel plants. A hundred thousand families have a saner, happier, more civilized home life every week for his breadth of vision and strength in pushing the matter against the inertia with which all proposals for change are met." [15] It was Kellogg who arranged with Samuel M. Lindsay, a prominent Columbia University economist, for Dickson to read his paper on liability insurance before the American Academy of Political and Social Sciences in 1911.[16]

Returning to Montclair from his first summer on the farm, Dickson met with Kellogg and apparently expressed an interest in writing on labor reform in the steel industry. Kellogg wrote to Walter Hines Page, editor of *World's Work*, and Robert U. Johnson, editor of *Century*, to suggest that they might be able to obtain articles by Dickson for their publications.[17] Although nothing came from these efforts, Dickson continued to work with the reformer in the hours battle.

It was Charles M. Cabot of Boston, however, the owner of a handful of shares of U.S. Steel stock, who between 1911 and his death in 1915 led the attack against the twelve-hour shift at U.S. Steel and its subsidiaries. Cabot did not fight alone, of course. Dickson, Fitch, Kellogg, and Lindsay, among others, assisted in various ways. But Cabot, by raising embarrassing questions at annual stockholders' meetings of U.S. Steel, attracted the headlines that publicized the crusade and helped to arouse interest in the cause.

Cabot was drawn into the contest by an editorial in *Colliers* magazine that critized U.S. Steel as an exploitive landlord. Although his holdings in the corporation were miniscule, Cabot assumed that he was personally responsible for the use made of property belonging to him. Writing to the corporation regarding the charges, he received assurances that they were false. When he called upon *Colliers* to retract, the editor put Cabot in contact with the Pittsburgh Survey people and Elizabeth Crowell, whose article had led to the editorial. Her evidence convinced Cabot that the corporation had lied to him. One thing led to another, and Cabot vowed to enlighten his fellow-stockholders and to stir them to action.

Cabot, with Fitch and Kellogg, called on Judge Gary to discuss working conditions in U.S. Steel plants about the time of Dickson's showdown with Gary over the Sunday labor issues in March 1910. Cabot proposed to hire Fitch at his own expense to prepare a report on wages, working hours, the speed up of work, and similar matters and to mail it to the corporation's shareholders. To the surprise of the reformers, Gary agreed to cooperate and promised to supply the names and addresses of stockholders.

When Fitch completed his study, Cabot decided not only to mail it to stockholders but to make it public as well. In March 1911 the study appeared as an article in *American Magazine* under the title, "Old age at Forty." The piece reported on the topics agreed to by Gary but also touched on the issue of the unionization of steelworkers. Gary charged that he had been imposed upon, both because the study took up topics he had not approved and because it had been issued publicly without his knowledge or consent. Under the circumstances, he refused to supply the names and addresses of shareholders as promised.[18]

At the annual meeting a few weeks later, Cabot offered a resolution calling upon the officers of the corporation to appoint a special shareholders' committee to investigate and report on the charges raised in Fitch's article. Gary protested that every opportunity had already been extended to Cabot and Fitch for their investigation. Because Fitch's article seemed "partisan, very unfair and unreasonable," and because he questioned Cabot's motives, Gary said that he had refused to assist in distributing the report to stockholders. As usual, Gary had absolute control over the meeting. Holding the proxy for nearly five million shares (as against a few thousand held by others present), the chairman could simply have smothered Cabot's resolution. But members of the press were present, and Gary, who had always insisted that U.S. Steel had nothing to hide, voted to hold the investigation.[19]

The stockholders' committee that Gary named was headed by Stuyvesant Fish, former president of the Illinois Central Railroad, and four other businessmen, including Charles L. Taylor, administrator of the Carnegie Relief Fund and Dickson's close personal friend. The committee did not meet until October and accomplished little before

January 1912 when Taylor induced the group to hire William H. Matthews, a Pittsburgh social worker, to organize and conduct the investigation. Dissatisfied with the committee's initial draft report, Taylor sent it to Dickson for comment. As the report stood, Dickson told his friend, it was a whitewash. If presented and accepted in that form, Dickson warned, he would feel compelled to attack it in the press. Taylor should either "insist on putting some teeth in it or decline to sign it." [20] The report was rewritten, strengthened, and printed for distribution at the annual meeting on April 15, 1912.

Meanwhile Cabot had instituted a law suit against the corporation to obtain the names and addresses of shareholders. That issue was settled out of court by compromise: Fitch was to rewrite his report, limiting it to the matter of hours of labor, Gary was to supply the names and addresses of fifteen thousand shareholders and Cabot, at his own expense, would mail out the report. In a covering letter, Cabot asked shareholders to express their views on the subject in letters to Judge Gary, "or better still . . . in person at the annual meeting in April." [21] Gary, too, wrote the shareholders, characterizing Fitch's report as the injudicious work of too-ardent an advocate of reform. The corporation, he insisted, was "not indifferent or self-satisfied" as to working conditions and was making improvements "as fast as practicable." The workers did not favor a change that would reduce their incomes by one-third, he pointed out, and business conditions barred any increase in wage rates at the time.[22]

At the meeting on April 15, Gary again held absolute sway, representing in proxy 4,952,757 shares of stock against 2,214 shares held by proxy or in person by ten others, including Cabot with 26 shares. Both the Fish Committee's report and a compilation of letters received from shareholders in response to Cabot's letter of March 26 were distributed to those present.[23] Fish read his committee's report to the assembly. It commended the corporation for its efforts to halt Sunday labor, but called for an absolute end to that practice "no matter what alleged difficulties in operation may seem to hinder." The "tendency" of any official "to disregard the spirit or the letter of such order should be sufficient cause for removal from service," the report suggested.[24]

Turning to the issue of hours, the committee reported that nearly 26 percent of all workmen employed by the coporation and between 50 and 60 percent of those operating blast furnaces, open hearth furnaces, and rolling mills worked the twelve-hour shift. Although the committee conceded that improved machinery had made work easier in recent years and that twelve-hour men had idle periods during the shift, it still believed that "a twelve hour day of labor, followed continuously by any group of men for any considerable number of years means a decreasing of the efficiency and lessening of the vigor and virility of such men." The committee softened the impact of its observation by expressing doubt that any single employer could afford to inaugurate shorter hours unless a similar policy was adopted by all employers in the industry.[25]

The balance of the report was highly favorable to U.S. Steel's other labor policies. The committee defended the piecework system and expressed doubt that foremen or superintendents who received bonuses for outstanding production records were driving their men excessively. Such incentives benefited both employer and employees, guarding against "that dead level of wages regardless of the ambition, the resourcefulness, the efficiency of the individual concerned." Indeed, piecework was "but a part of that spirit of contest and competition" that was "characteristic of American life."[26]

The suppression of labor unions was justified on the grounds that those organizations in the past had proven "irresponsible and incapable of self-control" and had advocated theories and practices "believed by many to be fallacious." [27] The report ended with praise for the corporation's Safety Department, its voluntary accident relief plan, its pension system, its improved sanitation and welfare programs, and its stock distribution scheme for employees. In sum, U.S. Steel had much to be proud of in its treatment of workmen. Sunday labor and the twelve-hour day, however, were abuses that should be ended.[28]

By April 12, only ninety shareholders of the fifteen thousand who received Cabot's letter and Fitch's report had responded. Five percent of the replies were ambiguous. Sixty percent supported Gary's position.

> I am perfectly satisfied to leave this matter with you and your board of directors.

> There is no demand or wish on the part of the steel workers for less hours.

> It will be a happy day for the business men, who are the only true friends the laboring men have, when a lot of these social fanatics are placed in lunatic asylums, and the muckrakers, labor agitators, and the grafters are put in jail.

> If Mr. Cabot don't want to be a party to the conditions he might sell his stock and thus relieve his conscience.

> Twelve or fourteen hours of work each day has never been known to kill.

> Where you benefit one man with an eight hour day you'll curse a dozen of his fellows with more temptations and extravagances.

> In my opinion the men injure themselves much more by dissipation than by work and are usually better off when they work every day. Then they escape 'blue Monday.'

> Iron and steel workmen, as a rule, are a very healthy, long-lived body of men. They suffer no physical exhaustion, either from over exertion or over heating. They work hard when they work, but the intervals between working periods are so long as to give them all needed rest.

Thirty-five percent of the respondents favored reducing hours, 13 percent even if it meant reduced dividends.

> If it were possible to have the men work eight, instead of twelve hours a day, I would be quite willing to receive 5% instead of 7%.

> If we can afford to write off for new constructions and improvements thirty or forty million dollars a year we can afford to be just and generous to our employees.

> If our corporation cannot earn dividends big enough for the majority of its stockholders without this injustice to men and their families, then better have a new set of stockholders.

> That the twelve hour day should persist no matter in how limited a number of cases, in an enterprise from which I derive money, gives me a sense of personal shame.

> The ultimate responsibility for bad conditions rests upon us, if we require of you only profits and not just conditions of labor. If such conditions of labor require a reduction in our dividends, in God's name reduce them![29]

Gary observed that many of those who favored cutting dividends to reduce hours of labor were "women or clergymen" who had "the least opportunity of knowing the actual conditions surrounding the mills."[30]

When Cabot was given an opportunity to reply, he suggested that all responses of stockholders had not been reported and that, as he remembered them, they had divided about evenly for and against him. He had no criticism of the Fish Committee's report, he said, but he had hoped that it would offer resolutions for carrying its recommendations into effect. Fearful that otherwise "things might go along about as they have been," he moved that Gary appoint a six-man committee "to consider the feasibility of introducing three shifts in all continuous processes," and, if feasible, to report to the next annual meeting a plan for putting such a schedule into operation.[31] Fish objected. His committee's report did call for "steps to be taken now." Moreover, by law the corporation had to be run by its elected officers, not by the shareholders at annual "town-hall" meetings. Cabot withdrew his resolution, expressing complete faith in both Fish and Gary and leaving it to the directors to take "such action" as they deemed advisable.[32]

Gary assured the assembled shareholders that the management of U.S. Steel conceived it to be their "first obligation to consider and determine questions relating to the welfare of our employees." The corporation needed "no magazine article, nor any resolution from any stockholder to spur us on in our endeavor to promote the welfare of the employees of the Corporation." At the same time, the management welcomed all evidences of stockholder interest in what the firm was doing and was pleased that shareholders called the management "to an account from time to time in regard to our stewardship." Personally, Gary added, he was not certain that twelve hours of work a day was "always a bad thing for the employees," especially when many "come to us insisting that they should have the right to work twelve hours a day in order to earn a little more money." As a boy on the farm he had worked more than twelve hours a day for years without harm, and at present the highest officers of the corporation "on the average work more than twelve hours a day." Nonetheless, Gary assured his listeners, the management would give its "best thought and most careful attention to all of the questions" raised at the meeting.[33]

A few weeks after the meeting, Judge Gary announced that the finance committee of the board of directors had set up a subcommittee consisting of himself, President Farrell, and Percival Roberts, a director of the corporation, "to consider what, if any, arrangement with a view to reducing the twelve-hour day" would be "reasonable, just and practicable." [34] Reporting at the 1913 annual meeting, the subcommittee noted that "only about 25%" of the corporation's employees now worked the twelve-hour shift. Moreover, U.S. Steel on its own could not abandon the twelve-hour day; unless its competitors cut hours simultaneously, steelworkers who preferred to work longer hours would quit just as they had when the corporation moved from the seven-day to the six-day workweek. Charles Cabot promptly offered a resolution that the corporation "enlist the co-operation of the steel manufacturers of the United States in establishing the eight-hour day in continuous twenty-four hour processes." Judge Gary exercised his block of proxies to table that motion.

Fitch, reporting on the meeting for *Survey*, noted that between 50 and 60 percent of the corporation's steelworkers still labored for twelve hours. The subcommittee had arrived at its lower figure only by adding in all of the firm's employees, including clerical workers, miners, railway workmen, and others. He also observed that the loss of men when the company abolished the seven-day workweek resulted from forcing the men to bear the full cost of the reform. Their weekly wages were simply reduced one-seventh.[35]

Between 1911 and 1914, *Survey* magazine kept the drums beating for hours reform, especially in the steel industry. Other journals, picking up the theme, held the demand before the public until the United States entered World War I. Not all the articles were the work of professional reformers or literary men. For example, R. A. Bull, president of Commonwealth Steel where the eight-hour shift had been instituted, attacked the longer work day. The public utterances of "some captains of industry" regarding work in the mills amused him because of their frequent misstatements of fact. Possibly ignorance of actual conditions accounted for their errors, he suggested. "You cannot expect any man to give you the best that is in him when you keep him employed without intermission for 12 hours per day, seven days per week, at work making a heavy demand upon his mental and physical powers, under conditions of high temperatures such as obtain on a furnace floor. To expect the best results under such circumstances is folly." Bull reported on a study made of the relative efficiency of workmen in his plant before and after the change to the eight-hour shift. "It will be observed," he commented on the statistics, "that the differences in most cases are slight, but the pleasing and important feature is that the essential ones are in favor of the short shift." Although he restricted himself to the efficiency factor, Bull did not wish to evade the humanitarian issue. "I do not hesitate to state my belief in the absolute injustice, humanely speaking, of the 12-hour shift on the furnace platform," he declared, and he added the hope that he would not be branded a socialist for so expressing himself.[36]

Dickson, whose previous labors on behalf of reform had been within the corporation or behind the scenes, publicly joined in the battle for shorter hours in January 1914. In an article in *Survey*, entitled "Can American Steel Plants Afford an Eight-Hour Turn?" he insisted that the reform was immediately practicable and implied that U.S. Steel should lead the way. If the federal courts (in the antitrust case then in progress) permitted the corporation to continue operating under its current structure, he said, the conditions that it set for its employees "must ultimately prevail in the entire steel industry." He endorsed the 1912 declaration of the Fish Committee that prolonged employment on the twelve-hour shift resulted in decreased efficiency, vigor, and virility of the workers. Indeed, he added, the majority of them did become "old men at forty."

To the corporation's plea that some workers opposed a shorter workday or shorter workweek, Dickson replied that the objections came chiefly "from that migratory class of laborers whose sole aim is quickly to accumulate some money and return to Europe, and who, in order to do so, are willing to live and work under conditions which are physically, mentally and morally debilitating." To allow this group to fix the work standards for American citizens, he declared, was not reasonable.

Increased efficiency from less-exhausted workmen, he argued, would offset any increase in costs. In the "remote contingency" that costs did rise, they should be passed along to the consumer in the form of higher prices for steel. After all, Dickson concluded, "the principal business of each generation" was not to produce cheap goods or to insure large profits to investors, but "to live normal human lives and to so maintain living conditions that succeeding generations may not be handicapped keeping the same standards." [37]

In April 1914, Dickson sought to enlist Andrew Carnegie in the battle for hours reform. Over the years Dickson had faithfully attended dinners of the Carnegie Veterans Association and frequently had written humorous verses, song parodies, and skits for the amusement of Carnegie and his guests. The Old Scot, apparently impressed by Dickson's skill with the pen, one winter day in early 1914 surprised the younger man by inviting him to play golf on his private links north of New York City. Although Carnegie had said that other friends would be present, it turned out that Dickson was the only guest. After "puttering around" the links "for an hour or so," the two men went to Carnegie's cottage and warmed themselves by the fire. Carnegie then retired for a nap, after which luncheon was served. Finally the "old boss" revealed the purpose of his invitation; he wanted Dickson to write an article on the Carnegie-Frick controversy that would justify Carnegie before the public. For several hours he worked on a memorandum, listing points that Dickson was to cover—notably Frick's handling of the Homestead Strike and Carnegie's later "eviction of Mr. Frick from the partnership." [38]

As might be expected, Dickson was fascinated by the story but soon concluded that what Carnegie proposed was unwise. Saying nothing, Dickson returned home and studied the notes at length. Although convinced that it would be "a monumental blunder" to write the article, he hesitated to make "a flat refusal." Instead he temporized, avoiding Carnegie for nearly three months. During the interval he drafted a lengthy letter, explaining his position but also trying to win Carnegie over to more positive projects. Concluding that it would be cowardly to mail the letter, Dickson made an appointment with the steelmaster and read the letter to him in person.[39]

In his letter Dickson traced his own long association with the steel industry from the day he started at Homestead through the struggle he had waged for hours reform as first vice-president of U.S. Steel. It was "the one great satisfaction of my life," he said, to have been "instrumental in striking the shackles from the American steel worker and have enabled him once a week, at least, to walk out under God's free heaven and breathe the air of a free man."

This satisfaction was mingled with bitterness, however, because to accomplish the reform he had had to antagonize not only Judge Gary, "who never forgave me for forcing the issue," but also "some of my dearest friends—men to whom I owed much of my success in life." All had come up together through the Carnegie Steel Company, which, Dickson charged, was "the most persistent offender in this practical enslavement of the workmen." Even today, he added, "while I believe the Corporation officials are sincere in their efforts to make the reform effective, the Carnegie Steel Company is the most difficult to hold in line." Although some Carnegie veterans at U.S. Steel had helped him, "certainly Schwab, Dinkey, Kerr and others were either strongly opposed or at least failed to support" him in the battle.

Having won the campaign against the seven-day workweek in spite of the opposition, Dickson noted, he recently had taken up the question of the twelve-hour workday and the right of workingmen to organize. Quoting at length from Carnegie's published views on these questions, Dickson observed that he had been "unable to avoid the conclusion" that critics would acknowledge Carnegie's sincerity in uttering these ideas and in putting them into practice as long as he was active head of his firm. The "later history of the Company," however, had been "directly opposed" to those beliefs. Particularly was this true of recognition of the right of workers to organize. "As a matter of fact, employees of our companies were not permitted to form organizations." Similarly, the twelve-hour workday, though once abolished by Carnegie, was "now with few exceptions the rule."

Instead of inviting "what must be a bitter personal controversy over the past," Dickson urged Carnegie to consider "the advisability of dealing with these problems as they exist today." Carnegie occupied a unique position and still had the "power to make history in the steel industry." His extensive holdings of U.S. Steel securities gave him "the moral, if not the legal right, to a voice in shaping its policy on this supremely important question of its relation to the human element in its organization." With his "immense influence," Carnegie could "clinch the six-day work-week in the industry" and "demand the change from twelve to eight hours in all continuous industries." On these subjects there was "no excuse for further delay."

Dickson admitted that recognition of the right of workers to organize was a debatable subject "due to the deplorable lack of real leadership in the unions as at present constituted." Nevertheless, he believed that the "present repressive policy" of U.S. Steel, "which in this particular matter adopted the Carnegie policy," was defensible only "as a war measure" and should be "abandoned without further delay." [40] Carnegie listened to the reading without comment. When Dickson finished, Carnegie looked up, changed the subject, and never again spoke of the matter. Dickson's frankness apparently did not lessen Carnegie's regard for him, but neither did it convince Carnegie to enter the ranks of the reformers.[41]

The most remarkable feature of the letter—from the standpoint of Dickson's evolving labor views—was its call for recognition of the right of workers to organize. Exactly when and why Dickson first stopped opposing the organization of workmen is uncertain. Perhaps his reading of Carnegie's pre-Homestead articles on labor, which included explicit recognition of the right of workers to form unions and to bargain with their employers through elected representatives, influenced him. At any event, Dickson's advocacy of worker organization in April 1914 put him well in advance of most of the nation's business leaders and certainly of those in the steel industry. Even Rockefeller's celebrated employee representation plan for the Colorado Fuel & Iron Company, the first in the industry, did not come into being until January 1915.[42]

Even as Dickson labored to win Carnegie's support for hours reform, Judge Gary was preparing a counterthrust against the eight-hour drive. He told the annual meeting of the corporation on April 20, 1914, that in wrestling with the question the officers had gone about as far as they could. They were now "confronted with the fact that the employees themselves objected to the reduction of hours, because necessarily it would result in a reduction of their wages." U.S. Steel already paid higher hourly rates than its competitors and could not afford additional raises to ease a transition from the twelve-hour to the eight-hour system.[43]

The Judge then flashed his trump card. Over fifty thousand employees of U.S. Steel also were stockholders in the firm, he declared, and their interests had to be consulted. As it happened, several employee-shareholders had been selected by their fellow employee-shareholders to be present to represent them at this meeting. Why and how these men happened to be selected to attend this particular meeting and none before or after, and how Gary knew they were present—if he did not arrange their coming—was not disclosed. Gary implied that he did not know what the men would say, but, when given their chance to speak, they expressed complete faith that U.S. Steel was doing everything in its power to improve the lot of employees.[44]

The relationship of the workers and the corporation, one employee-stockholder declared, "is what I call A Number One." [45] Another said that in his opinion "the twelve-hour man as a rule is not a hard-working man at all. I am inclined to think it would be conservative to say that there is one-third of the time as a rule that the twelve-hour man does not actually work; and in most cases his work is not heavy work while he is working." The rolling mill where he himself labored twelve hours a day, he continued, "is one of the greatest places to labor you have ever seen. You do not hear anybody complain about it." It was his contention that the men wanted to work the longer hours. He personally favored ending the twelve-hour shift, he confessed, but did not see "why it should be rushed through all at once and the entire system destroyed." [46]

Another stockholding employee praised the "good, clean, cold water" that the company piped directly to the workers, "the clean condition inside the mill," the disinfected toilets and shining windows, the cooling system, the emergency hospital, the pension system, and the stock purchasing plan. Now fifty-two years old, he had worked for various steel companies in Kentucky, Ohio, and Pennsylvania and was content to stay where he was because he knew he could not "get better conditions in any other plant in the country than what I have in the Steel Corporation." [47] A blooming mill worker boasted that "the conditions in our mill are perfect. We are all satisfied and contented. And any time we have any complaint we go right to the superintendent and we always get a fair hearing, and he sees that everything is right. The foreman don't drive anybody. He never says a word hardly to anybody." [48]

The climax to the proceedings had come a bit earlier when one employee-stockholder suggested that "we ought to throw away all hammers and quit our knocking and quit our kicking and look up and thank God for this great corporation that is taking care of humanity in this large way." He then asked those assembled to bow their heads as he prayed,

> Our God, we thank Thee for these good men who have taken such an interest in the welfare of humanity, and rejoice in our hearts for the blessed privilege of looking up to Thee in thankfulness for it. And we would pray for the man who works and all who keep us together as one great family and organization. In Thy name we ask it, Amen.[49]

Swept along by the spirit of the proceedings, one stockholder said that his father, who had been in the iron business, had wanted his son to have "an easy job" and so had "made a doctor" out of him. "I am going to apply for a job in the Steel Company," he declared.[50] The meeting must have seemed to Gary an unqualified success. Even Charles Cabot appears to have been won over. Early in the proceedings the Judge detailed all the actions taken by the corporation in response to Cabot's questions, noting the heavy costs involved. Cabot responded by praising Gary and the corporation for the "splendid work done" in improving conditions for employees.[51] Dickson jotted on his copy of the published minutes of the meeting that he was not much impressed by the views of Judge Gary's "led horses."

The counterattack on the eight-hour movement was broader than the staged annual meeting of 1914. Both U.S. Steel and the industry at large turned to public relations programs to improve the image of steel manufacturing. *Leslie's Weekly*, reportedly under the control of J. P. Morgan, ran articles praising the industry, and in 1915 Arundel Cotter, with the cooperation of Gary and the corporation, published his *Authentic History of the United States Steel Corporation*. The tone of Cotter's work is indicated by the subtitle of a later edition: *A Corporation With a Soul*. More important in quieting the critics of the industry were the ongoing labor reforms at U.S. Steel. Ending the twelve-hour shift was not among those reforms, however.[52]

In fact it would be World War I that would halt and then reverse the drive for hours reform in the steel industry. Ultimately the reformers lost when the industry appeared to yield to their demands. Under strong governmental pressures, the steel companies adopted the "basic eight-hour day" on October 1, 1918. However, since the ever-increasing demand for steel and the growing shortage of labor made an actual eight-hour day impossible, the workmen simply received overtime pay for all time they worked beyond eight hours. The arrangement muffled the cries of reformers, increased the number of men working the twelve-hour shift, undercut Dickson's hard-won battle against unnecessary Sunday labor, and quietly expired at the end of the war when the overtime pay (but not the long hours or Sunday labor) was dropped.[53] In 1911, slightly more than 45,000 (approximately 26 percent) of U.S. Steel's employees worked the twelve-hour shift. By October 1920 the figure had risen to about 85,000 (or 32 percent) of the men.[54] Writing on January 20, 1920, Gary admitted that "during the war, at the urgent request by government officials for larger production, there was considerable continuous seven-day service in some of the departments." [55] Although exact statistics are not available, one student of the question concluded that the seven-day week at the corporation "assumed large proportions in part of 1916 and all of 1917, 1918, and 1919," though it was "probable . . . that conditions never became as bad as they were in 1910." [56]

* * *

The question might well be asked, why did Gary and his chief associates resist hours reform with such determination? Between 1902 and 1911 they had instituted a broad range of worker welfare programs at considerable expense with only minimal internal debate. Did hours reforms somehow differ from those other reforms, and, if so, how? Over the years Judge Gary offered a number of explanations: (1) the reform was being promoted by well-meaning people from outside the industry who mistakenly believed that they knew better than the responsible officers of the corporation what was best for U.S. Steel's employees; (2) twelve hours of labor in itself was not harmful to workers; the chief officers of the corporation, for example, regularly worked twelve or more hours a day; (3) the work of the twelve-hour men was not continuous; there were frequent periods for rest and relaxation; (4) the workmen, whose wishes should carry greater weight than the demands of outsiders, opposed moving to the eight-hour shift; (5) workers would accept the short shift only if they could earn as much in eight hours as they were currently earning in twelve hours. The costs of a raise was greater than the corporation or industry could afford.[57]

Most of Gary's arguments do not stand up under examination. "Outside" reformers did actively promote hours reform, but Dickson was not an outsider, and men like Cabot, Fitch, the Fish Committee, and others were at least as well informed on the hours question as Gary himself. Moreover, it required little expertise to realize that the hours worked by steelworkers was excessive by the standards of the day. The overwhelming majority of American workers labored fewer than sixty hours a week. Outside of steel, less than ten hours a day was the norm.[58] At the same time, none of the other conditions of labor prevailing in the industry (such as pay, cleanliness, ease of task, opportunity for advancement, and the like) offset the long hours. And while it was possibly true that officers of the corporation often put in more than twelve hours a day, their situation was hardly comparable to that of the men in the mills. Officers were free to take time off without penalty, the nature of their work was varied and tended to enlist their interest, it frequently involved travel, meeting with other men of affairs, and making important decisions. Their jobs brought them influence, power, and wealth. Unlike the men in the shops, Gary and his fellow officers enjoyed lengthy summer vacations at full pay and lived lives of great comfort if not outright luxury.

The twelve-hour shift, however viewed, was neither easy nor relaxing. Steelmaking was hard, dirty, exhausting work. The twelve-hour men were giving fully half of their life (and three-quarters of their waking time) each day to their work and earning a scant livelihood. There were breaks in their long workshifts, but they were brief and the men could not leave the shops or otherwise spend the time as they wished. For the most part they rested in preparation for the next arduous round of labor.

As for the men "preferring" the long shift, reformers were quick to point out that without an increase in wage rates, the change to an eight-hour shift would reduce the workers' pay by one-third. Most employees and their families hovered too near the subsistence line to survive so drastic a cut. Obviously it was necessity, not preference, that led the men to cling to the twelve-hour shift.

These more frivolous excuses aside, two of Gary's arguments seem to offer the most likely reasons for his resistance to hours reform: managerial prerogative and costs. By law, control and management of U.S. Steel and its subsidiaries properly rested with Gary and his chief associates. Certainly he would not have surrendered any of his authority to others, nor would he bow to outside pressures. Either, to him, would have been an indication of weakness on his part or an admission that his critics were right. Had Gary yielded on the hours questions when Dickson first raised the issue, the reform could have been put through without his appearing to have surrendered to outsiders. The longer Gary delayed, however, the more the outsiders clamored, and the more difficult a backdown became.

In the end costs may have been the most important consideration behind Judge Gary's rejection of hours reform even as he was actively supporting worker welfare programs. In 1912, the first full year of operation of the complete welfare program—including "welfare," "sanitation," company housing, accident prevention, accident relief, and pensions (beyond those financed by the earnings of the permanent pension trust fund)—the cost to U.S. Steel totaled nearly $5 million. That sum equalled about 3 percent of what the corporation spent that year on wages and salaries.[59] Even $5 million was misleadingly high. Many of the worker welfare programs paid back much, if not most, of the money put into them. As already noted, many items charged to "welfare" or "sanitation" were housekeeping costs that the corporation would have spent with or without the welfare program. Money spent by U.S. Steel on company housing came back to the corporation in the form of rent or repayments, with interest, on home loans.

According to the corporation's assistant general solicitor, if the cost of accidents borne by U.S. Steel in 1906 (the year the safety program began) had remained unchanged through 1911, the corporation would have paid out $3,862,000 *more* than the $2,500,000 expended in that period on accident prevention. The cost of the accident relief program brought no direct return to the corporation. It made possible, however, a reasonably accurate prediction of accident costs each year and largely freed U.S. Steel from unpredictable jury awards to injured workmen who otherwise might have sued the corporation.[60] Only the expenditures on pensions—except for the good will they generated—were a dead loss to the corporation.

By contrast, hours reform would have involved a very considerable increase in costs. Exactly how much was difficult for reformers from outside the firm to determine because they did not have access to essential data—the number of men working the twelve-hour shift, the number still working seven days a week, hourly rates of pay, and the like. Within the corporation there appears to have been a curious lack of interest in the subject. If Gary, for example, had estimates made of the possible cost of changing to an eight-hour shift, they have not yet leaked to the public. Even Dickson, who had complete access to all essential information and who argued for hours reforms on both humanitarian and efficiency grounds, seems never to have calculated the cost of adopting the eight-hour shift.

From data supplied by the Fish Committee's report in 1912, an estimate of costs can be made. Slightly more than a quarter of U.S. Steel's employees—45,284 men—worked the twelve-hour shift. Changing to the eight-hour shift, while maintaining continuous operations, would have required hiring 22,642 additional men. Assuming no loss in daily income for those who previously worked twelve hours a day, and equal pay for the new employees, the wage cost of the changeover would have amounted to approximately $18.6 million. This figure is reached by multiplying the number of new employees by the average annual wage of all U.S. Steel employees for 1912 ($822). Under the twelve-hour shift system, total wages and salaries at U.S. Steel was equal to 26.1 percent of the total receipts of the corporation. The changeover, as calculated, would raise total wage and salary expenditures to 29.1 percent of receipts. This 3 percent increase could have been met in any of several ways—by increasing prices of steel, by reducing dividends on stock slightly more than a third, or by taking that amount from the undivided surplus funds of the corporation which, in 1911, amounted to more than $135 million.[61]

It is possible, indeed probable, that if the income of continuous-operation employees had increased by 50 percent, it would have been necessary to make adjustments in the earnings of the remaining three-quarters of U.S. Steel's employees, thus adding even more to the overall cost of the changeover. On the other hand, it is probable that some of the added costs would have been offset by the increased efficiency of men working the shorter shift.

Given the scale of U.S. Steel's operations, a 3 percent—or even 5 or 6 percent—rise in the cost of production might seem a small price to pay to benefit so large a number of workmen. An unspoken consideration through the debates may well have been Gary's concern for the impact of such costs on the sale of steel abroad. Prior to the twentieth century, the United States had exported little steel. Moving into foreign markets had been an important factor in the creation of the trust in 1901. By 1911, the corporation exported 80 percent of all steel leaving the United States—an amount, by weight, equal to over 14 percent of U.S. Steel's total output. Adding 3 or more percent to the price of steel might well have had a serious impact on those sales.[62] Welfare reform clearly cost far less than hours reform.

* * *

The same flood of war contracts from Europe that swamped the demands of Dickson and others for hours reform brought Dickson an opportunity to return to a major managerial post within the steel industry. Although American steel producers strained their plants to capacity and wrung from their men ever greater output and longer hours of work, they could not keep abreast of the orders for guns and munitions pouring in from Europe. American financiers, eager to gather in the lush profits dangling temptingly before them, soon loosed a flood of fresh capital into war production. With the formation of new enterprises, the demand went out for experienced men to run them.

In March 1915, Corey invited Dickson to accompany him to Europe. To whom the former steelmen talked and what understandings they reached, they kept strictly to themselves. Dickson did not even preserve any details of the meetings in his personal papers. Upon their return, Corey entered into a syndicate of manufacturers (including Ambrose Monell, a one-time metallurgist for Carnegie and now president of International Nickel, and Samuel Vauclain, president of Baldwin Locomotive) and leading New York investors (Percy A. Rockefeller, M. Hartley Dodge, and Frank A. Vanderlip, president of National City Bank) to create a major new steel combine. The firm, Midvale Steel & Ordnance, was a holding company chartered under Delaware law with an initial capital of $100,000,000. Leading bankers and a few armaments manufacturers—Albert H. Wiggin, president of Chase National Bank; Charles H. Sabin, president of Guaranty Trust Company; Joseph W. Harriman, president of Harriman National

Bank; Frederick W. Allen of Lee, Higginson & Company; Samuel F. Pryor, vice-president of the Metallic Cartridge Company; and W. P. Barba, vice-president of the Midvale Steel Company (one of the plants bought up by the combine)—joined with syndicate members to constitute the firm's board of directors. Alva C. Dinkey, president of Carnegie Steel, was lured away from U.S. Steel's managerial staff to head the new organization, while Dickson became the firm's vice-president and treasurer. These men, too, sat on the board of directors presided over by Corey.[63]

In its quest for properties, the syndicate settled on three firms near Philadelphia: the Midvale Steel Company of Nicetown, Worth Brothers at Coatesville, and the Remington Arms Company at Eddystone. According to the *Wall Street Journal*, Corey and his associates "jumped for Midvale, which for thirty years had been building big guns and armor plate for the United States Government." It also manufactured a variety of steel products "for commercial uses," including marine engines, castings, tool and automobile steel, axles, and steel-tired wheels. The plant, covering fifty-two acres, consisted of crucible and open hearth furnaces, forge shops, a projectile department, and large machine shops. With 5500 employees, "a half-dozen men in every department that knew every wheel to be turned," and located in "the best technical labor market in the country," Midvale offered "great possibilities of expansion both in plant and organization." [64] Because the retiring president had one daughter married to an Englishman and another to a German subject, Midvale had remained strictly neutral in the war, refusing to sell arms to either side. The new managers, however, expected to "engage extensively in the manufacture of munitions for the European belligerents." [65]

Worth Brothers, a successful competitor of Carnegie Steel for over a generation, had an "up-to-date" plant with two modern 500-ton blast furnaces "built upon the latest United States Steel Corporation plans." "[L]aid out on lines for broad expansion," the company occupied but a third of its 700 acre site. Specializing in plate and tube production, its plate capacity was second only to that of U.S. Steel. The third property, Remington Arms Company, rounded out the new combine's facilities for supplying "everything, from armor plates, big guns and shells to rifles." Chiefly a manufacturer of small arms, Remington at the time it was purchased held a contract for two million Enfield rifles on which it expected to realize ten dollars apiece, enough to reimburse the syndicate in full for its purchase price of the plant.[66]

In all, the three properties cost Corey and his associates $60.5 million. The difference between that sum and the $75 million in issued stock (believed to represent the subscriptions paid in by the syndicate members) was to serve as the combine's working capital.[67] The financing of the new firm promptly gave rise to speculation as to the purpose behind it. That the syndicate had been able to coax Dinkey away from Carnegie Steel led one commentator to suggest that there must be a big stock-jobbing scheme afoot. "As steel capitalizations go," the *New York Evening Post* noted, Midvale's was "altogether excessive." The *Wall Street Journal*, on the other hand, predicted a bright future for Midvale. The company had no funded debt and its stock represented full value in plants (less water not exceeding $5 million, promotion commissions, and bonuses). With its plants located at tidewater, where business would be concentrated for the next six years or so, Midvale would avoid heavy land transportation charges.[68] No trouble was expected from the antitrust laws—the syndicate had been careful to avoid merging competing firms. At the same time, relations with other steel giants would be harmonious; Dinkey was the brother-in-law of Charles M. Schwab, the president of Bethlehem, and, according to Corey, Midvale had "no policy of destructive competition in view." It was "in business simply to make money." [69]

The syndicate seemed especially anxious to dispel rumors that the combine was merely speculative or a war venture. The *Wall Street Journal* reported that Midvale was "not a financial promotion" but "a cash enterprise to fill orders." It was "generally understood in the steel trade" that the combine had "comprehensive plans for expansion into commercial lines of steel." Dinkey, Corey, Dickson, and Monell, "of the old Carnegie school" and "prominent bankers," would not have ventured into a scheme limited to "profits on war orders alone." Midvale's promoters insisted that Midvale was a "peace stock," not a "war stock." "The Old Carnegie crowd, which will operate Midvale Steel . . . backed by the most powerful financial interests in the world," must have foreseen much more than was obvious to the public, the *Wall Street Journal* observed. The United States would be doubling its navy within five years, the war-time wear and tear on European navies would create large orders for repair parts, and merchant marines would be expanding greatly.[70]

The rumors about stock-jobbing persisted, nonetheless, until Midvale made further acquisitions. In December 1915, the syndicate took up an option held since October on 300 million tons of nickel-rich Cuban iron ore. Fifty million tons would have been sufficient to maintain current rates of operation; 300 million tons gave Midvale a reserve of ore one-third as great as that of U.S. Steel.[71]

A second coup followed in February 1916 when Kuhn, Loeb & Company of New York decided against financing a merger of the Lackawanna and Youngstown Steel Companies with the Cambria plant at Johnstown, Pennsylvania. Learning of Cambria's availability from a friend of Carnegie Steel days, Dickson urged Corey to purchase the firm. Assisted in the negotiations by the Philadelphia agent of J. P. Morgan & Company, Midvale suddenly closed the deal, thereby adding to its own output of 650,000 tons of steel ingots per year an additional 1.25 million tons. With Cambria's ingot capacity of between 1.5 million and 1.75 million tons, Midvale's total capacity now ranked second to U.S. Steel's capacity of approximately 20 million tons. A producer of rails, structural shapes, plates, agricultural steel, spring steel, wire, and freight cars, Cambria manufactured no munitions at the time of its purchase. The "general assumption," *Nation* reported, was that Midvale acquired the Johnstown firm "as a safeguard against the return of peace when orders for munitions will dwindle away." [72]

Whatever Midvale's long range prospects, its immediate undertakings were to supply munitions to the Allies. By the end of October 1915, Midvale already had war contracts nearly equal to those of Bethlehem: almost $600 million worth. One order for 4 million rifles (on which it would earn $15 profits each) would net Midvale $60 million.[73] At the end of its first year of operations, Corey reported "earnings in excess of 30 per cent on the stock, exclusive of profits from a contract for 3,000,000 rifles." With all plants operating at full capacity, Midvale's earnings exceeded $2.5 million per month.[74] In 1917, net earnings reached almost $70 million. A 3 percent quarterly dividend on the stock was declared in early 1917 and continued through the war years. Despite rising wages and costs and heavy corporation and excess profits taxes after America entered the war, by the end of 1918 Midvale had accumulated a surplus in excess of $53.5 million.[75]

But the war brought problems as well as profits to the steel industry, and Dickson found it necessary to deal with labor shortages, governmental price controls, and taxes. In the initial wave of enthusiasm over entering the war, Dickson feared that too many steelworkers might enlist, thereby depleting the labor supply. Writing to Secretary of War Newton D. Baker, Dickson asked if President Wilson could not issue a proclamation emphasizing the vital necessity to national defense "that men in certain selected industries should remain at their posts, at least until the national authorities have decided that their services will be more valuable to their country elsewhere." [76]

As Midvale strained to keep up with the demands of the government for war material, Dickson drafted a "100 Per Cent Pledge" designed to spur both the officers and employees of the company. Officers were asked to promise "the prompt production and delivery of the largest possible quantity of material in our departments that is or shall be required by the United States Government for the necessities of itself and its Allies," and to agree that "all other lines of our business shall be subordinate to this pledge." Each employee was to acknowledge as his "patriotic duty" reporting for work every day and "while on duty to devote every energy to the efficient performance of his particular work." Only with the full cooperation of all could Midvale produce the steel needed for "the prosecution of this great *WAR IN DEFENSE OF CIVILIZATION.*" With "our sons and brothers . . . baring their breasts to the foe on the battlefields of France, ready to make the supreme sacrifice for our beloved country," it was "both our patriotic duty and our sacred privilege to do our part to see that not one of these lives is unnecessarily sacrificed because of lack of equipment which can be furnished through our labor." [77]

Attempts of the government to hold down steel prices without at the same time controlling the costs of labor, raw materials, and other factors of production drew Dickson's wrath. "Manufacturers and businessmen generally," he wrote in a letter to the editor of *Iron Age*, "should be allowed to make liberal profits, so long as their business is being conducted under fair competitive conditions." Once the profits were made, the government should take by taxation "their fair proportion of whatever may be necessary to provide the funds for the prosecution of the war.[78]

The new and higher federal taxes occasioned by the war were not wholly wise or fair, Dickson believed. For example, he favored a general sales tax on all production, to be passed along from raw materials producers to middle men to ultimate consumers. Such a tax would be fair, universal in its incidence, and easy to collect.[79] Exemptions from the income tax, he argued, should be "so low as to bring the great majority of our citizens within its operation and thus give them a vital interest in Government financing." The excess profits tax should be based on a clear definition of capital invested and should exempt a reasonable return on such capital—perhaps 10 percent—after providing for depreciation both ordinary and special. Special depreciation should be allowed for "extra hazardous capital investments" and for investments in plants and equipment the value of which would be impaired by the ending of the war. To assure both fairness and sufficient flexibility to be practical, Dickson recommended the establishment of a permanent tribunal to rule on such matters and a governmentally prescribed uniform accounting system for all corporations.

Dickson further believed that the costs of the war should be divided into "irretrievable expenses" and "investments." The former, which would include the cost of assembling, equipping, and maintaining the armed forces, should be paid out of current taxes. Investment expenditures, including loans to foreign governments, loans to farmers, advances to manufacturers, and moneys spent on such permanent improvements as docks, harbors, waterways, railroads, and the like should be financed by bond issues.[80] It was Dickson's contention that future generations "who will reap the benefit of our sacrifices" should carry "a more equitable share of the burden" than had thus far been proposed.[81]

Once the United States entered the war, the nation was flooded with what H. L. Mencken later called "the bilge of idealism." Many hardheaded businessmen cynically turned the rhetoric of patriotism to their own advantage. Dickson's frequent references to patriotism and sacrifice differed from the general run in that he seems to have been sincere. Ever an idealist, Dickson wholeheartedly subscribed to the extravagances of Wilsonian slogans as true expressions of his own and the nation's highest principles. He filled his diary (apparently kept for his own purposes and not for posterity) with notes of concern at the progress in the battle between freedom and militarism. He bore the increased tax burden on his personal income as a necessary price of freedom and willingly participated in bond rallies.[82] He watched as Charles, his only son, enlisted in the Naval Aviation Corps, then among the more dangerous services. When the boy "actually started to fly," Dickson afterward recorded, "I considered him as a sacrifice in this great War in defense of Civilization and now that it is practically ended I think of him as one come back from the dead." [83] One of his daughters, Emma, volunteered for work near the front in France as a canteen girl. When at last the Armistice came, the Dickson household was awakened at 4 A.M. by shouting and whistles. "After breakfast," Dickson wrote in his diary, "we sang patriotic songs and hymns, closing with the doxology. We then took three cars and went out among the neighbors to start some enthusiasm." [84]

Given his strong sentiments about the war, Dickson was shocked and upset by the attitudes of some of his associates, especially Corey.

> Long talk on Democracy vs. Autocracy with WEC he strongly justifying the latter. He expressed himself in the strongest terms in favor of autocratic government and the exploitation of the lower classes by those who by *any* means physical or intellectual or financial are able to raise themselves above the 'common herd.' He admires and justifies the Germans and reiterates his

regret that America did not cast in her lot with the Kaiser and 'divide the world between them.' I tried to reason against this barbarous creed, stating that while recognizing the ills and weaknesses of democracy, I believed the remedy lay in more democracy, which of course presupposes the education of every person in that democracy into the status of a thinking unit. I also pointed out the fact that the basic wrong in the theory of autocracy lay in the monopolizing of natural resources and especially land, by the ruling class thus reducing the vast majority of people to the level of serfs. My argument fell on deaf ears, however. A few weeks ago I had a similar argument with Dinkey and Neale while dining with Charles at the Waldorf, both of them expressing similar views to these of WEC although not so entirely frank. I consider it a matter of the most serious regret that three men so closely identified with our company should be so permeated by undemocratic ideals.[85]

Dickson may not have accurately interpreted the views of his associates, or they may have enjoyed pulling the leg of their overly idealistic colleague. Dickson, who knew them well, however, and regarded Corey as his closest and dearest friend, took their remarks at face value.[86]

During the war Dickson once again had the opportunity to press for labor reforms in the steel industry—this time in the area of collective bargaining. Whatever Corey and Dinkey thought about Germany and autocracy, they had already demonstrated beyond doubt that they saw no place for any "democracy," however little, in the relationships of employers and employees.

CHAPTER

5

Experiment with "Industrial Democracy"

"INDUSTRIAL DEMOCRACY," as Dickson liked to call it, came to the Midvale Steel & Ordnance Company with unseemly haste during the last two weeks of September 1918. On the afternoon of Thursday, September 19, a special meeting of the board of directors authorized the officers to establish a "comprehensive system of collective bargaining." [1] That same night Dickson began work on the plan that eventually would be adopted. He spent the morning of Saturday the twenty-first in conference with the chief officers of the company (Corey, Dinkey, Slick, and Neale) and the afternoon and next morning discussing the plan with managers of Midvale's plant at Coatesville. After dispatching revised copies of the plan and instructions to the Cambria plant at Johnstown, Dickson returned to Philadelphia to talk with the managers of the Nicetown works. About eleven on Sunday morning, the twenty-third, notices were posted in the plants, informing the workers that Midvale now recognized their right to bargain collectively and inviting them to attend meetings in the mills the next day to elect representatives.[2]

At each of the Monday meetings in the various plants, an official of the company read a statement prepared by Dickson. Midvale was inviting its employees "to co-operate . . . in formulating a system of collective bargaining, which will be of mutual advantage to the company and all its employees." An employees' association, "entirely in the hands of men elected by the votes of their associates," was to be set up. To get underway, the workers were first to elect a chairman of their meeting and a secretary to record what was done. They were then to elect representatives to a plant committee. The plant committee, in turn, was to meet immediately and select from among themselves a specified number of delegates to attend a conference at the Philadelphia headquarters of the company on Wednesday the twenty-fifth.

There the delegates from the three plants, meeting with company officials, would draw up a plan that, when completed, "would be submitted to the workers for ratification." [3]

At Philadelphia on September 25, Dickson welcomed thirteen worker-delegates (five each from Nicetown and Johnstown and three from Coatesville) "as equal partners" with himself, E. E. Slick, and the general superintendents of the three Midvale plants in formulating a plan of employee representation. The step that Midvale was taking in democratizing relations with its employees, Dickson told the group, was "so fundamental and far-reaching" that it, like the American Revolution, the Declaration of Independence, and the Civil War, would mark "an epoch worthy to be commemorated by future generations." After further remarks in the same vein, Dickson called upon the workmen to select from among themselves a chairman and a secretary for the meeting. That accomplished, Dickson, "for the purpose of facilitating the work of the conference," presented his "tentative draft of a proposed plan." The small group devoted the balance of Wednesday and most of Thursday to discussing and voting on details of the proposal, one by one. The group then unanimously adopted the plan "as finally amended by them." [4]

Contrary to what had been promised at the initial meetings on Sunday the twenty-second, the plan was not submitted to the rank and file for ratification. Instead, plant committee meetings were called on Friday the twenty-seventh to adopt the plan that had been completed in Philadelphia only the previous afternoon. Proceedings went smoothest at Nicetown where the plant committee met at 5:00 P.M. and "unanimously adopted" the plan, presumably in time to get home for supper. The minutes of the meeting noted no discussion whatever. At the Coatesville meeting, copies of the plan were distributed to plant committee members to study, and another session was called for September 30. At the second meeting the plan was "unanimously accepted." According to the minutes, "individual discussions were then heard, each member giving his opinion; all being highly in favor of the adoption." The Johnstown plant committee meeting scrutinized the plan more carefully. Each part was discussed and voted on separately. A motion was then made, and seconded, to adopt the plan, "subject to the ratification of the employees of the Cambria Steel Company." For reasons not explained in the minutes, the mover and seconder of that motion withdrew the proviso that approval be subject to ratification by the employees, and the plant committee then adopted the plan as submitted.[5]

The plan went into effect on October 1, only twelve days after the

Midvale board of directors first authorized collective bargaining. As will be seen, the speed with which the plan was adopted resulted far less from enthusiasm at the prospect of employee representation than from a desire on the part of the company to avoid an even less desirable alternative.

* * *

Dickson's efforts to improve hours of labor and other working conditions in the mills grew out of his boyhood experiences as a steelworker. His interest in the right of workers to organize and bargain collectively came much later and was always tempered by his distrust of regular trade union leaders. Unfavorable experiences with unions during his formative years left Dickson with an antiunion bias that years in the management of the industry had only reinforced.

Among Dickson's earliest memories (predating his father's fall from wealth) were tales of violence committed by the Molly Maguires. He first heard about the Mollies while visiting the home of a schoolmate whose father was an impoverished coal miner. The stories took on significance when the superintendent of a mine adjoining one of his father's properties was assassinated, allegedly by a Molly, and when John Dickson received threatening notes.[6] As already related, Billy Dickson and his older brother Robert continued to work at Homestead during the strike there in 1882. Robert was beaten by a band of patrolling strikers, and Billy, after the strike, was permanently scarred when union men deliberately threw a rail against his leg.[7]

By the time of the great Homestead Strike in 1892, Dickson had become a clerk in the Pittsburgh offices of Carnegie Steel. His brother Robert was a foreman in the mills and another brother, Tom, worked in the foundry and belonged to the union. On the day of the battle between the strikers and two bargeloads of Pinkerton detectives who were coming by river to occupy the plant, Dickson and a group of clerks from the Pittsburgh office watched from the opposite shore.[8] The horrors of that bloody conflict burned vividly in Dickson's memory through the years. Initially sympathetic to the company, in time he came to believe that the excesses of the men were the result of the brutalizing conditions under which they labored. And so not only zeal motivated Dickson to end unjust and barbaric conditions in the mills. He also hoped that future confrontations and bloody encounters could be avoided.

By 1914 Dickson's concern began to range beyond hours reform and improved work conditions. The impact of the deadly monotony of mindless factory work on laborers, the growing gulf between management and labor, and the attendant antagonism and lack of understanding between managers and workers all troubled him. Where the character of work stunted the laborer "mentally, morally and physically," Dickson could only suggest locating factories away from crowded cities and reducing the number of hours spent tending machines. With such reforms a workman could devote his increased leisure "to a more rational method of living:" to "the cultivation of vegetables and fruits for the sustenance of the bodies of his family, and of flowers for their souls." Something along the line of Henry George's Single Tax was also essential, he believed, if land and other natural resources were to be freed from the grasp of monopolies and made more generally available.[9]

To close the gap between management and labor, Dickson advocated the general adoption of stock ownership by employees. Like Perkins of U.S. Steel, he believed that stock ownership would improve worker motivation and tend to increase production. Dickson also took up Perkins's suggestion that employees should "elect fellow workmen as directors" because the worker-directors would "have an opportunity to come in touch" with the problems of management, and "could, in turn, transmit some appreciation of them to the rank and file." [10]

Dickson seems first to have advocated the organizing of workers for purposes of collective bargaining early in 1914, shortly before writing to Andrew Carnegie to enlist his support behind hours reform at U.S. Steel. Thereafter his thinking advanced rapidly. In spite of improving working conditions, Dickson told a meeting of the Friendly Sons of St. Patrick in Montclair in January 1915, "the right of the workmen to collective bargaining" could not be "permanently suppressed if our republican form of government" was "to endure." The organization of labor was a necessity. "It is the only logical answer to organized capital. . . . The individual workman, dependent on his own strength and resources, cannot hope to bargain on equal terms with the corporation. *If he cannot do so, he is no longer a freeman but a serf*, and the serf has no place in the future of America."

Unions, once purged of false leaders and other evils, would be legalized, he predicted, and employers by law would be prevented from discriminating against union members. "If I were a workman I would belong to a union," he declared. "Even as I am, I would be tempted to join one if I were eligible." Dickson could imagine "no more glorious" career than leading a union, "inspired by the spirit of service, to help solve the complex problems which lie before us." [11] Still in retirement at the time, Dickson clearly ached for a leadership role—in industry or even as a champion of the workingman.

How soon, how hard, and how far Dickson pushed his new views once he entered the management of Midvale Steel & Ordnance is not recorded. That he discussed the question with his superiors as often as opportunity afforded cannot be doubted, given Dickson's temperament and zeal. The reactions of Corey, Dinkey, and the others are equally certain. It was not until the United States entered World War I and the Wilson administration began vigorously to court organized labor's support for the war effort that Dickson would have the opportunity to put his theories to the test.

* * *

The war-born need for an uninterrupted flow of reasonably priced steel inevitably drew the federal government into such matters as pricing and labor relations. Steel manufacturers, who regarded these areas as their exclusive preserve, both resented and resisted governmental intrusion. As intervention became more insistent, the steelmasters united behind Judge Gary, in effect making him their intermediary for dealing with Washington. Combining footdragging with skillful negotiating, Gary worked out a modus operandi with the War Industries Board under which the problem of steel pricing was resolved on terms highly satisfactory to the industry.[12] Settling the labor question proved much more difficult.

Once the United States entered the war, President Wilson's hitherto ambiguous stance on labor questions gave way to open wooing. He began by tacitly accepting "American Labor's Position in Peace and War," a declaration drawn up on the eve of the war by a conference of labor leaders headed by Samuel Gompers, president of the American Federation of Labor. In return for its full support in the event of war, labor insisted that capital not be permitted to profit at the expense of workingmen, that wages and work conditions "conform to the principles of human welfare and justice," and that organized labor be recognized as spokesman for workers, representing them on all war agencies. During the first year of American participation in the war, the administration worked through Gompers and existing machinery of the Labor Department to avoid strikes and work stoppages.[13]

However, in April 1918, upon the recommendation of the War Labor Conference Board, Wilson assigned the task of handling disputes in war industries to a special agency, the National War Labor Board (NWLB). Cochaired by former President William Howard Taft (chosen by representatives of industry) and labor-lawyer Frank P. Walsh (chosen by representatives of labor), the board consisted of five industrial and five labor spokesmen. Having no enforcement powers of its own, the NWLB relied on the weight of public opinion or supportive action by the president to back up its awards.

The NWLB announced that it would adhere to the principles recommended by the War Labor Conference Board: strikes and lockouts were to be avoided, the right of workers to organize was not to be denied or abridged by employers, workers engaged in legitimate trade union activities were not to be discharged, where union shops existed they were to be continued and where the open shop was in effect it was not to be disturbed, the prevailing wage in an area was to govern wage settlements, and the right of all workmen—including common laborers—to a "living wage" was recognized.[14]

As might be expected, unions and companies interpreted the statement in accordance with their respective interests. When the unions began a drive to organize the steel industry, for example, they saw it as only the exercise of a legitimate trade union activity. The companies, of course, complained that the drive violated the pledge that wartime conditions would not be used to advance the position of either capital or labor at the expense of the other.

Warily, the NWLB sought a path between the obviously inconsistent principles it had adopted. That path, pleasing to neither unions nor companies, consisted of ordering the establishment of shop committees or employee representation plans in war production plants threatened with labor disputes. Such organizations made possible a kind of collective bargaining without forcing the companies to recognize or deal with regular trades unions. The NWLB was not the first to follow this course; as early as October 1917 the Shipbuilding Adjustment Board had ordered the creation of shop committees in shipyards that faced labor problems, and both the United States Railroad Administration and the United States Fuel Administration had followed similar policies. Once the NWLB adopted the shop committee idea, its decisions, one scholar has suggested, "were characterized by a steady and almost severe insistence that collective dealing [as opposed to collective bargaining with bona fide trades unions] be established as a normal process in industry." The board, indeed, seemed to consider the establishment of shop committees more important than the resolution of immediate grievances. During 1918 and 1919, various federal agencies ordered the creation of 128 shop committees in plants engaged in war production. Of these the NWLB was responsible for 86.[15]

The issue of collective bargaining began to stir the steel industry early in 1918, even before the NWLB came into being. On February 14, Corey, chairman of Midvale's board of directors, surprised Dickson by asking him to confer with Bethlehem's president, Schwab, with regard to the "general question of the attitude of American employers to labor." Relations with labor, Corey suggested, might well have to be handled "somewhat differently" than in the past. "The world do move!" Dickson commented in his diary.[16] The outcome of the meeting apparently was not recorded, but for the next several months labor relations at Bethlehem and Midvale followed parallel courses. Organizers appeared at both firms in April 1918. That summer both companies were hauled before the NWLB for refusing to bargain collectively with their employees. By autumn both had set up employee representation plans in order to escape recognizing or dealing with regular unions.[17] The parallel may have been coincidental, but relations between Schwab, his brother-in-law Dinkey, Corey, and Dickson—all graduates of the Carnegie mills—had always been close.

Midvale first came under attack when the International Associaton of Machinists (IAM) began organizing its Nicetown plant. The success of the drive among the ten thousand employees is not certain. A union attorney's statement that the IAM represented some two thousand Midvale employees was probably an exaggeration. Concentrated among fifteen hundred employees in Machine Shop 7, the IAM was able to take out no more than six hundred men during a walkout in June.[18] Whatever the number of members, on April 30, William A. Kelton, the IAM's local business manager, addressed a letter to Midvale officials, urging acceptance of a draft agreement that he enclosed. The principal changes proposed by the agreement were an increase in wages to cover recent rises in the cost of living and adoption of a basic eight-hour day. "You may feel some reluctance in entering into an agreement with a labor union," Kelton observed, "but I am sure that if you will adopt this method of dealing with your employees you will find that harmony among the workmen will be promoted and production stimulated." Kelton notified the United States Department of Labor of his action.[19]

Midvale officials made no reply. When a Labor Department mediator arrived on the scene to offer his services, Henry D. Booth, general superintendent at Nicetown, informed him that there was nothing to conciliate. A committee of employees belong to the IAM that called on Booth on June 14 to discuss the agreement were told that it was contrary to company policy to bargain collectively or to consult with employee committees. When the men returned to their benches, company officials told them that if they were dissatisfied they should leave. As the commmitteemen packed their tools to depart, between three hundred forty and five hundred forty others joined them.

Learning of the troubles from the IAM, Navy and Labor Department spokesmen pressured Midvale's president, A. C. Dinkey, to yield. When he refused, the navy asked the workmen—on patriotic grounds—to return to their jobs on June 17. The men agreed, only to find that the company would not take them back. A telegram from the assistant secretary of the navy assured the men of that department's interest in the matter and guaranteed that substantial justice would be done. The navy then called on the NWLB to intervene, and, at the insistence of board examiners, Midvale rehired two hundred forty of the men who had walked out on June 14. The company refused, however, to discharge any employees hired in the meantime to create jobs for one hundred (the IAM claimed from two hundred fifty to three hundred) others who had quit work.[20]

As the controversy continued, Dickson concluded that the time had arrived for proposing "industrial democracy" at Midvale. At a meeting of the board of directors on August 7, he introduced a resolution recognizing and welcoming the principle of collective bargaining and declaring that the "right of workingmen to organize to protect their rights is beyond question." To determine employee attitudes, the resolution proposed meetings in the shops, conducted "entirely by the employees, without any interference, direct or indirect, by any superintendent, foreman, or other person having authority over them." By secret ballot the men would vote yes or no on the proposition, "Do you desire, as an employee of this works, to form a permanent organization with your fellow employees, so as to be in a position to bargain collectively with the company on all matters pertaining to wage rates and all other working conditions?" [21] The board took no action. Another turn of the screw was needed to bring Midvale officials to accept so "radical" a step.

Midvale's case came before a section of the NWLB on August 28. There, John F. Perkins, an industrial representative on the board who had received a lengthy letter from Dickson regarding the situation, summarized the controversy and proposed an award.[22] He noted that the chief demands of the men were for shop committees and collective bargaining, adoption of a basic eight-hour workday, modification of the bonus (or piecework) system, and elimination of the Midvale Beneficial Association—a supposedly voluntary mutual assistance society to which most Midvale employees contributed. The workmen also complained of discrimination against union members. Perkins supported the company's position, finding, he said, "no discontent or dissatisfaction of any moment among the men." Except for individual complaints such as were bound to arise in any plant the size of Nicetown, the discontent that existed was caused by the activities of union organizers "and did not spontaneously grow up in the plant." The NWLB had been created "to preserve industrial peace and maintain production," he pointed out, and Midvale had been "remarkably successful in securing and maintaining production of war material." Furthermore, the firm's "organization and methods of dealing with its men" had been "worked out after years of careful study and adjustment." He proposed, therefore, that the NWLB recommend to the company the adoption of "a proper system of representative committees of their own men," and that all other matters be left to the company and its employees to resolve, the board declining to take any action regarding them. Unwilling to accept Perkins's recommendations, the board decided instead to confer with Dinkey.[23]

Meanwhile, new pressures had brought the labor problem to a crisis in the steel industry at large. In May 1918, President Wilson had set up the War Labor Policies Board (WLPB), with Professor Felix Frankfurter of the Harvard Law School as its secretary. The board was to coordinate labor policies for the various government departments and war agencies. It soon became evident to steel executives that their authority over their employees was being challenged by the government on two fronts. While the NWLB insisted that companies bargain collectively with their employees whenever disputes arose, Frankfurter and the WLPB were pushing for adoption of the eight-hour workday as a first step in stablizing wages in the industry. The WLPB hoped the move would halt labor shortages caused by men changing jobs in quest of higher pay.

In July, Frankfurter invited Judge Gary to name a committee of steel managers to help the WLPB in its work. Suspicious as to what might be afoot, Gary delayed his reply for over a week. Finally he wrote that he would be able to cooperate more fully if Frankfurter or one of his deputies—when next in New York—would call on him and explain what the board had in mind. Frankfurter's response confirmed Gary's suspicions. The WLPB hoped to arrange a conference of labor and management representatives to discuss a wide range of labor complaints, including overly long hours, that if not resolved could lead to interruptions in steel production. Frankfurter's letter made clear the intention of the WLPB to involve itself on a grand scale in the problems of steel labor.

To meet the threat, Gary summoned a special meeting of 150 steel executives at the Waldorf-Astoria on August 28. There the industry's leaders decided that the committee previously appointed by Gary to work with the government on steel prices should also represent the industry in all dealings with the WLPB.[24] As in the matter of pricing, Gary in effect would speak for all, and his policy would be to stall any reforms as long as possible. At Frankfurter's insistence, a meeting was held on September 20. Five days later Gary announced that as of October 1, 1918, the eight-hour "basic day" would be substituted for the twelve-hour shift. The wartime shortage of labor, however, prevented the actual adoption of an eight-hour workday. Men continued to work in the mills twelve hours as before, but were paid time-and-a-half for all time over eight hours. The arrangement satisfied neither Gary nor the reformers. To the Judge the arrangement was nothing more than a scheme for raising pay. Reformers chafed that men were still working intolerably long hours.[25]

Meanwhile, Midvale's time for stalling ran out. On September 16, company officials received a telegram from Secretary of the Navy Josephus Daniels, who was "surprised and somewhat disturbed" that Midvale had not yet submitted its differences with its employees to the NWLB for settlement. The fact that in June the men had returned to work "in good faith pending arbitration," made "acquiescence" by the company "a matter of honor, as well as common sense." "Please inform the Taft-Walsh Board immediately of your entire willingness to submit your side of the case, the men already having submitted theirs. I shall expect a telegram to that effect," he concluded, "as the matter is urgent and serious." [26]

The telegram, Dickson observed, precipitated a "crisis" that "must end in the acceptance of some form of collective bargaining." He had been "advocating this," he noted, "in a quiet way somewhat along the lines of the Colorado system." (On August 29, Dickson had conferred with officials of the Colorado Fuel & Iron Company with regard to the employee representation plan they had set up in January 1916, following the Ludlow Massacre.) "E. E. Slick [a vice-president of Midvale and president of the Cambria division at Johnstown] is a reactionary and W. E. C. [Corey] and A. C. D. [Dinkey] are on the fence. The question seems to me to be 'Shall we jump or wait to be pushed?' " [27]

While Dinkey and Midvale counsel T. L. Chadbourne conferred with Secretary Daniels in Washington, Dickson called on Judge Gary (as president of the American Iron & Steel Institute) to present a memorandum on Midvale's problem. The company was being pressured by the government to submit to the War Labor Board "a demand made by a Labor Organization with which we have never had any dealings" and to abide by the ruling of the board. Midvale's directors would be acting on that demand within a few days. Company officials suspected that the move was "the beginning of a well-organized campaign to force the Unionization" of all its plants and eventually of the industry as a whole. The scheme no doubt originated with the American Federation of Labor and its affiliates, but "Government officials closely identified with the Administration" were effectively supporting the drive in violation of the spirit of President Wilson's announced policy that wartime conditions were not to be used to change the relation of any company to its employees. Protest by a single firm would carry little weight, Dickson noted. The "only way" to save the industry from being "completely dominated by irresponsible labor leaders" would be for Gary, as spokesman for the entire industry, to discuss the matter frankly with President Wilson. Dickson added that Midvale was considering and would "probably adopt" a system of collective bargaining similar to that of the Colorado Fuel & Iron Company. "Adoption of such a plan," he declared, would "furnish a complete answer to objections which have been raised as to the present system." [28]

The next morning, Dickson, Corey, and the general superintendents of the various Midvale plants decided to recommend to the board of directors that the controversy at Nicetown be referred to the NWLB and that an employee representation plan be introduced for the company at large. Although neither Corey nor Dinkey would advocate collective bargaining, they allowed Dickson to propose his scheme to the board. That afternoon the directors yielded to Secretary Daniels's demand and, to the surprise of Corey, Dinkey, and Dickson, approved setting up a system of collective bargaining. Corey, in notifying Daniels that the dispute was being submitted to the NWLB, noted that the secretary would probably be interested to learn that Midvale, which "for some time" had been studying the matter, was about to invite its employees to join in consideration of a plan of collective bargaining.[29]

Except for Dickson, Midvale officers had yielded to employee representation only with the greatest reluctance. Now, however, they raced to put the scheme into operation before the NWLB could act on the Nicetown controversy, which involved recognition of the machinists' union. That same day, September 19, Dickson drafted a notice for Dinkey to sign, informing the employees that the company recognized "the right of wage-earners to bargain collectively with their employers," and inviting them to meet with company officers in their respective plants. At those meetings they would consider and, if practicable, adopt "a plan of representation by the employees, which shall be thoroughly democratic and entirely free from interference by the companies, or any official or agent thereof." [30] By October 1, as already noted, the plan of employee representation was drafted, adopted by the plant committees, and rushed into operation.

In its final form the plan called for the annual election of representatives to a plant conference committee on the basis of one for every 300 men in each plant division. The resulting committees consisted of 56 members at Johnstown, 25 at Coatesville, and 35 at Nicetown. These representatives, in turn, were to elect one representative for every 3,000 employees to serve on a company-wide general conference committee. A representative was required to have worked in the plant for at least a year and would lose his position if he ceased to be an employee or was promoted to a salaried position. A representative could be recalled if two-thirds of the men he represented signed a petition against him.

Those elected were to act as spokesmen for the men "in all matters pertaining to conditions of employment, the adjustment of differences, and all other matters affecting the relation of the employees to the Company." When a foreman or division superintendent failed to settle a workman's grievance, it could be appealed to the local plant conference committee which would discuss it with the general superintendent. If not satisfactorily adjusted, further appeal could be taken to a joint meeting of the general conference committee and general superintendents of all Midvale plants, to the president of the company, or, if need be, to binding arbitration.

Under the plan the company's right to hire was unlimited, and it was free to suspend work in any department because of a "lack of orders or for any other legitimate business reason." Although such a suspension could be made without warning, it was "the duty of the officers to give as much advance notice as practicable."

The right of the company to suspend or discharge individual employees was spelled out. A man could be discharged immediately and without notice for disloyalty by act or utterance to the U.S. government, any offense against the criminal law of the state, assault upon or attempt to injure another person, wanton destruction of property, refusal to obey a reasonable order of his superior, or intoxication while on duty. A man had to be warned at least once before being suspended or discharged for "carelessness," for failure to report for duty regularly and on time, or for "inefficiency, etc." A discharged man could demand that the cause "be clearly stated to him," and appeal could be taken to the plant general superintendent. The company promised not to require employees to purchase any goods or services from the company.

The capstone of the plan, Dickson believed, was the provision that all representatives from all Midvale plants were to meet four times a year with the officers of the company. At these sessions "all matters of general interest to both parties" could be discussed. It was at these meetings that Dickson hoped to narrow the gap and end the antagonism between employers and employees.[31]

Dickson was pleased with his work. The plan marked "a distinct epoch in the history of American business," he believed. "Without question all other steel companies, including the United States Steel Corporation," would be "forced to follow our lead in this matter." He was "profoundly convinced" that "however disturbing" it might be initially, the proposed plan "for the first time in the history of industrialism" recognized "the principles of democracy as the only proper basis upon which an industrial system can be built which will receive the approval of that final Court of Last Resort,—Public Opinion." [32]

In his enthusiasm, Dickson considerably overstated both the novelty of the Midvale plan and the extent to which it gave workmen a real voice in determining the conditions under which they were employed. He later claimed to have weighed for several years the problem of how best workers could organize—or be organized—for purposes of collective bargaining. Yet when the Midvale board of directors approved of employee representation as a means to avoid recognizing the IAM, Dickson had no specific formulation in mind. He not only consulted with officials of Rockefeller's Colorado Fuel & Iron Company about their plan of employee representation, he also freely borrowed from it, paraphrasing its language and adjusting its specific provisions to the situation at Midvale. Although Midvale was apparently the first eastern steel plant to establish a company union, both Lukens and Bethlehem Steel adopted similar plans within a few days.[33]

Dickson and other Midvale officials, in introducing the plan, explicitly referred to it as a system for collective bargaining. The term did not appear in the text of the plan, however. Like the Rockefeller scheme, the Midvale plan might better have been referred to as a "representative system" of "communication and conference" between company officials and employees. While the plan spelled out election and grievance procedures in elaborate detail, listed specific offenses that could result in an employee being dismissed, and imposed no limits on what employee representatives could discuss with company officials, it did not suggest any of the areas that might or should be discussed. Also, there were no provisions for written contracts, and, unlike the Rockefeller plan, the Midvale plan designated no person or persons to carry into effect such decisions as might be reached by company officials and employee representatives.[34]

* * *

With its hastily formulated scheme of "industrial democracy" nominally in operation, Midvale appeared before a section of the NWLB to defend itself against the charges of the IAM. The union demanded a wage increase, establishment of the eight-hour workday, time-and-a-half for overtime, double pay for work on Sundays and holidays, reinstatement of "discharged" employees—with back pay for lost time, recognition by the company of the right of employees to join labor organizations of their choice, an end to compulsory membership in the Midvale Beneficial Association, acceptance by the company of collective bargaining "in all dealings with the workers," abolition of the recently adopted employee representation plan, equal pay for women engaged in the same work as men, discontinuance of the bonus (or piecework) system, payment of workers on company time, and retroactivity of the NWLB's award to June 18.

The company refused to raise wages or to hire men who had quit work on June 14 unless their services were actually needed. It also declined to break up the employee representation plan or to end the bonus system, and it urged the NWLB not to make its award retroactive. The other demands, the company insisted, were already in practice in one form or another.[35]

In their attack on the employee representation plan, IAM members charged that the program had been sprung without notice and rushed through before the workers understood what was happening. No time was allowed for employees to get together to talk over the idea. When men from Nicetown's Machine Shop 7 (where the IAM was strongest) asked for three or four days to think over the proposition, they were told "that it was all planned and had to go through now." Two days later "the job was done." One elected representative admitted to IAM business manager Kelton that he "didn't know any thing at all about what was going on . . . and didn't understand anything about it." A second representative, unfamiliar with collective bargaining, asked an IAM member to explain what the term meant.[36]

Even more telling was the charge that managers at Nicetown had interfered in the selection of representatives. As a result, the majority of those chosen were "foremen." Although the men actually were "sub-foremen," or gang leaders and wage earners rather than salaried employees, they did exercise quasi-managerial powers. At the initial meeting in Machine Shop 7, the superintendent announced that nominees for the department's four representatives to the plant committee had to have been employed at the plant for at least a year and were not to be salaried foremen. Workers "rather forcibly expressed the opinion" that foremen of any kind (even those on wages) should not be eligible. Shop 7 elected only workmen.

Discovering at the plant-wide committee meeting that eighteen of the thirty-five representatives were nonsalaried foremen, the representatives from Shop 7 promptly resigned. At the subsequent election to replace them, only 450 of 1,300 employees participated (between 600 and 700 had taken part in the earlier election) in choosing three workmen and one foreman as representatives. The Nicetown plant committee, in turn, elected five members to go to Philadelphia to help draw up the plan. Four were nonsalaried foremen—one having held the position for four years, two for ten years, and one for twenty years.[37]

Under questioning by the IAM's attorney, Dickson came close to conceding the charges against the Midvale plan. Employee representation "as an abstract proposition," he said, had been in his mind "for many years." He had undertaken to put it in "definite form . . . as I recall it, about April." The plan took "final form early in September" and was approved by the board of directors on September 19. As for the source of the plan, it had "largely . . . evolved" from his own experience of over thirty-seven years in the steel industry.[38]

Turning to the matter of the elected representatives, the IAM lawyer asked Dickson if it would be a "fair restriction" to bar "all foremen and others who have power to exercise authority over the employees" from election as representatives. Dickson agreed. The attorney pressed on. How would Dickson have felt at the meeting in Philadelphia had he known that the majority of representatives elected at Nicetown and the majority of the group meeting to adopt the plan were foremen? Dickson conceded that this did not meet his ideal "to the extent of 100%." At the same time, he added, "if the men decided that these men who were foremen, and at the same time wage earners, were their natural leaders, they had a right, as the plan then stood, to elect them."

Asked to react to the fact that salaried foremen at Nicetown "were busy in the plant suggesting names" of men to elect following the call for the initial meetings, Dickson said he regarded it as "an exercise of over zealousness" by those men.

Q. Would you have gone so far as to have considered that the men would have a justification in believing that, while the company was pretending to adopt an equitable plan, it was at the same time endeavoring to keep such close control over the plan that it would not be effective and beneficial to the men?

A. I might say that I can see how such an impression might be given. . . .

Q. Knowing, as you must know, that the National War Labor Board, has heretofore insisted that corporations largely engaged in war work production must make some arrangments providing for . . . collective bargaining as between the company and the working men, and knowing that this plan was presented after a hearing of the controversy now under investigation had taken place before a section of the War Labor Board, would you not think that the employees would have the right, and be justified in the suspicion that this plan was brought forward for the purpose of avoiding a plan that might be later suggested in these hearings before the War Labor Board?

A. Such a supposition on the part of employees who had lost confidence in the management might be natural. I can only answer that by stating that the evidence which has been brought before you this morning, shows, I think, conclusively, that the matter had been under consideration for months, prior to any difficulty between the company and its employees. . . .

Q. It appears in this case, that on the 30th day of April, 1918, the business representative of the IA of M in the City of Philadelphia, representing some 2,000 employees of the Midvale plant, submitted a form of agreement to the superintendent of your plant, together with a letter, calling attention to the facts that I have suggested and stated in my question. Do you consider that an effort towards collective bargaining?

A. Undoubtedly.

Q. After your many years of study, Mr. Dickson, what criticism have you to offer for that method of collective bargaining?

A. I would use the phrase that was once used by a distinguished President of the United States, in dealing with a problem, that it was a condition and not a theory that confronted us.

Q. We may be equally frank now, may we not, in writing into this record, that your company does not want to deal with organized labor, in the ordinary sense that we use that term?

A. Not as at present administered.

Q. If you deal with your men, by any method of collective bargaining, you insist that they must be a collection of individuals organized in your own plant?

A. For the present.

Q. What do you mean by that?

A. I mean that I hope that organized labor . . . will in the course of years make itself acceptable to employers.

Q. Have you ever thought that one of the easiest ways to bring that condition about would be by a close co-operation between the employers and these various organizations?

A. That is possible.

Q. Do you think that organized labor can ever reach the ideal state, so far as an ideal is possible, so long as employers compel them to fight for every condition that they are able to get?

A. There is a good deal in what you say, no doubt.

Q. You believe in the eight hour day?

A. I do.

Q. And you know, as a matter of fact, that the Manufacturers' Association, of this country, has continuously and determinedly fought that eight hour day, and compelled these men to strike, compelled them to resort to the strike, and the boycott, and everything else, in their effort to get it?

A. Undoubtedly there are reactionaries in the association to which you refer, and whom I condemn in the most emphatic way.

Q. Is the great issue today, between these groups of laboring men and groups of employers due to an utter lack of confidence upon both sides?

A. Undoubtedly.

Q. Do you think that that chasm of lack of confidence can be bridged by the intelligent men on both sides of that controversy coming together?

A. I do, and I think this [Dickson held up a copy of his plan] is one of the bridges. . . .

Q. As a matter of fact, that is not a bridge at all, it is a tunnel, isn't it?

A. I do not think so. . . .

Q. Have you not purposely, intentionally, adopted this plan, in order to try to initiate your ideal, or accomplish your ideal, without dealing . . . through organized labor?

A. Yes, without dealing with organized labor as it is at present administered; kindly underline those last words.[39]

* * *

As the NLWB pondered its decision from November to February, the Midvale Employee Representation Plan began to function. The first quarterly meeting of company officials with employee representatives took place in November at the Bellevue-Stratford Hotel in Philadelphia. The setting—with company officers seated behind a table on a dais and the representatives of the workmen clustered around tables below—and the remarks of the officers, revealed the essentially paternalistic tone of the meeting. President Dinkey told the men that there was "almost no difference" between employees, department superintendents, and other officials right up to himself. "Some of the best friends I have are workmen in the works," he declared. "The best friends I ever had have been workmen in the works." Every officer present, he observed, had started in the shops and moved forward to his present position. It was the men themselves, by the way they cooperated or failed to cooperate with a foreman, who determined whether or not he would advance. "It is his associates who really appoint him. The printed notice that is afterwards put out by the management is, after all, only a matter of form."

As for the meeting itself, its purpose, Dinkey said, was "to find out the best thing to do to make our work as a whole a real success." After all, he told the men, "we have no other excuse for living than to successfully perform the work that is placed in our care. We go before the public as good steel makers, good forge makers, good gun makers, good wire makers, and good plate makers. We work into these articles something of our own character."

Dickson, the driving force behind employee representation, had little more of a concrete nature to propose. Paraphrasing the Gettysburg Address, he portrayed the session as a great contest to prove whether or not American factory workers and their employers could run businesses democratically. Stressing the equal importance of capital, labor, and management, he urged those present to impress upon the men they represented "that it is of the utmost importance to our mutual interests" that they meet the company in the same "spirit of fairness" that the company was meeting them. "Do not," he pleaded, "let any outside influence inject discord into our relations at this critical time."

The chief business at this first session consisted of a discussion of substituting a comprehensive new insurance and pension program for the existing beneficial associations at the Johnstown and Nicetown plants. Under the new plan, employees and the company would jointly contribute to a fund out of which death, disability, sickness, and retirement benefits would be paid. The role of the worker representatives was largely passive. Dickson had begun work on the plan with company officials and insurance agents in mid-October and completed details of the plan following the meeting without further participation by worker representatives. Near the close of the session, worker spokesmen raised issues of their own, but because of the essentially local character of the points they raised, nothing of company-wide significance was decided.[40]

"About 200 present," Dickson wrote enthusiastically of the meeting. "Discussed plan of insurance, etc. sanitary measures, and other matters of general interest. Served lunch and dinner and dismissed about 10 PM after a very enjoyable and profitable day." [41] The thoughts that the worker representatives carried home apparently were not preserved.

From the very beginning, Dickson had proposed a company magazine to promote communication between Midvale's managers and workmen. The first—and only—issue of *The Steelworker*, put together entirely by Dickson, appeared in January 1919. A slick, twenty-four page publication, it contained little of obvious interest to the men in the shops. The cover carried a drawing of a noble-browed, square-jawed steelworker wielding a large hammer. Inside a "greeting" explained that the first issue was intended "to promote a mutual acquaintance" between company officers and workers. The chief officials of the company, Dickson commented, "in the course of years must be succeeded by younger men, and in all probability, if we are justified in reasoning from the past, the future Presidents, Vice Presidents, and

General Superintendents of our Companies are now, as young men, laying the foundations of their future success in the ranks of our workers." Four pages were given over to the minutes of the first quarterly session in Philadelphia, complete with Dinkey's and Dickson's opening addresses. Interspersed among the minutes, and taking up nine pages, were full-page pictures with biographic sketches of Corey, Dinkey, Dickson, Slick, and the other officials of Midvale. Seven pages of pictures of worker representatives (group pictures without biographic sketches) followed. As filler, Dickson used inspirational verses from his reading and his favorite poem, "Abou Ben-Adhem." The final page was devoted to a list of the directors and general officers of the company, the meeting time of stockholders, and the status of Midvale's capital stock and bonds as of November 30, 1918. Even this well-meant but pathetic endeavor to reach out to the men drew a negative response from Dickson's superiors. "It was to have been issued quarterly," he noted. "Corey and Dinkey got 'cold feet' and I was compelled to abandon the plan." [42]

In February 1919, the NWLB issued its award. Midvale was to establish the eight-hour day at Nicetown and to pay a specified minimum wage for each job. The board also appointed an examiner, John O'Brien, to determine whether the new system of collective bargaining was fair, and, if not, to order new elections to select representatives.[43] O'Brien interviewed Dickson and came away very impressed. "He is a man of big, liberal ideas regarding unionism, but says he disapproves of some of the acts of the leaders. He was against Gompers until the war came on; now he is a Gompers man. He thinks Gompers and the union men behaved gloriously during the war crisis." Dickson told O'Brien that the leaven of collective bargaining was spreading and that within ten years would be the rule in all plants. Profit sharing was also coming; "the day when a steel corporation, for instance, will produce forty millionaires and forty-thousand underpaid workmen—slaves—is gone forever in America." Dickson hoped that the Midvale plan would be amended to exclude all foremen from serving as shop committeemen. Asked if he saw the plan as a substitute for regular unions, Dickson replied, "No. I would be perfectly willing to have the union swallow it up, if the union could run it to the advantage of the men in the shop and in fairness to the company. My doctrine is not narrow. I am out for the general welfare—the greatest good to the greatest number and a square deal for all." [44]

O'Brien interviewed other company officials, union organizers, and hundreds of workers both in the shops and outside—at meetings, on the streets, in bars, and in their homes. On March 11, he wrote to his superior at the NWLB that he had been invited to meet with a group of workmen at Nicetown, only to find himself at an IAM meeting. There the men told him of "interference, intimidation and coercion on the part of foremen and representatives of the company." In one shop, just before the voting, workmen found notes pinned to their time cards urging the election of a particular candidate. Since company rules forbade workers from touching one another's time cards, the company itself seemed to have joined in the campaigning. O'Brien asked why the men had not told him this during his interviews with them in the shops. They replied that they were "afraid to speak to . . . any visitor freely; not knowing but that they might be speaking to an enemy in disguise." The examiner asked for documents to substantiate the charges and the men agreed to supply them. Convinced that the voting in that particular shop had been unfair, O'Brien was prepared to call for new balloting.

Three days later O'Brien concluded that he had been taken in. The union men had not delivered the "proofs" as promised, and a supporter of one of the candidates in the election confessed that he had pinned the campaign material to the time cards. In his final report on March 22, O'Brien ruled that the elections had been fair and that no new vote would be required. The failure of the Midvale plan to provide for an amending process had apparently been an oversight, he added, and would be corrected at the next quarterly meeting. In discussing the pros and cons of allowing nonsalaried foremen and gang leaders to serve as shop committeemen, O'Brien observed that these men did not have the power to hire or fire, and, being the most intelligent of the workers, were needed as representatives. At the Philadelphia meeting that drew up the original plan, O'Brien said, Dickson had asked the thirteen representatives whether foremen should represent workers. The thirteen, a majority of whom were themselves foremen, had seen no objection.

On the other hand, O'Brien noted, foremen were obviously beholden to the company and would be under temptation to see its point of view. Also, although they had no authority to discharge employees, their opinions carried great weight with their superiors. Under the circumstances, O'Brien recommended that the plan be amended to exclude foremen from nomination in future elections. Meanwhile, the NWLB should "make no order which would disturb the present Shop Committee," which was "an exceptionally efficient one." The committee had adjusted "numerous petty complaints and grievances" and committee members "individually and collectively" had assured O'Brien that collective bargaining was "working out to the general satisfaction of the men in the plant." The chief complaint that workmen made of the Midvale plan, O'Brien noted, was that foremen could serve on committees. Even so, except for Shop 7 when IAM sentiment remained strong, two-thirds of the hundreds of men that he had interviewed "had no fault to find" with "foremen" serving as committeemen. "While personally, all my sympathies are with organized labor," O'Brien declared, he could not and would not decide in favor of the IAM or any other union when the union was wrong. "The Machinists' Union is wrong in this case." As for the rest of the NWLB's February award, the examiner noted that Midvale had accepted it "in all respects" and was "complying with it." [45]

A portion of O'Brien's survey was later published—by whom is not clear. The document failed to do justice to the examiner's extensive interviewing of workmen. Only eleven of the Nicetown and sixty-five of the Johnstown interviews with laborers were mentioned. The published version, nonetheless, contained insights that O'Brien's letters and reports did not. For example, it revealed the split among top executives at Midvale with regard to the collective bargaining plan. Dickson and Nicetown General Superintendent Booth were enthusiastic. They admitted that union men ridiculed the plan and that some workmen were indifferent, but over all, collective bargaining had been completely successful. The "beauty of the plan," Dickson pointed out was that it relieved the shop management of "all the detracting annoyances" and released "50% of their time to their work." Booth agreed. Prior to adoption of the plan scarcely a day passed when he was not called on to deal with individual or group grievances. During the first 120 days that the plan was in operation, only three matters had come to him, "all others being adjusted before they got out of the department; in fact, the foreman is more on the alert to avoid the development of grievances in his department." Dickson suggested that the plan was "applicable to any class or character of employees, provided it was completely, intelligently and carefully introduced. Booth warned that it would not work where employers retained a "narrow, penurious, dominating, autocratic, shortsighted" relationship with their men. A manufacturer who "puts in this plan without careful preparation . . . only drives another peg for the union agitator to hang his hat on."

President Dinkey's secretary, on the other hand, knew "very little" about the plan except that it was Dickson's idea. At Johnstown, Vice-President Slick was "very reluctant to say anything whatever about the plan," it being "too new to be of interest to anyone yet." According to the assistant general superintendent at Johnstown, Slick and other executives there were "not very favorable" to the plan, seeing it as "aiding the American Federation of Labor" in that union's drive to organize the Johnstown works.

Worker reactions to the plan may have been less enthusiastic than O'Brien reported. Nicetown had cut employment from a wartime peak of nine thousand to four thousand. Open criticism of the company to outsiders by those still working could hardly be expected. At a mass meeting called by union organizers at Johnstown, O'Brien talked to about fifty employees. While they "would not condemn" the plan, they said it "did very little good" and "did not get them as much as a union could get them." Of the sixty-five individual workers for whom there were reported interviews, nearly half were evasive, non-commital, or afraid to answer, or were ignorant of, or indifferent to the plan. Twenty-seven favored the scheme; six opposed it. O'Brien's interviewing techniques, moreover, left much to be desired. He accepted as employees views, for example, statements made for them by wives, relatives, and other workmen. No attempt was made to probe such responses as "the company was always doing something nice for the help," or that the plan was "no d—— —— good. It is only a capitalist's 'bunk' to keep the men from organizing," or "Job all right, company all right. Soon no work. Everything all right. Work for ten years." O'Brien was not without prejudice. For example, responses were classified not according to the workman's reaction to the plan but by ethnic background—"American," "Colored," "German," "Irish," "Polish," "Slavish," "Serbian," "Hungarian," "Italian," and "Miscellaneous." One interviewee, the examiner observed, appeared to be "too stupid" to respond.[46]

* * *

By mid-March 1919, not only was the Midvale plan in place and functioning, it had won the endorsement of the NWLB. For Dickson, a great opportunity for revolutionizing the relationship of employers and employees now beckoned. "The only way out of this senseless conflict between capital and labor," he declared, "is for employers to realize that the day of industrial democracy has dawned and that the establishment of wage rates and other conditions of employment without representation is tyranny." [47] His more cynical colleagues saw the plan as having already served its purpose. It had satisfied federal officials who were insisting on some kind of collective bargaining, while it blocked the efforts of the IAM to expand its small toe-hold at the Nicetown plant into full recognition by the company.

But a further test of employee representation lay just over the horizon: the drive of organized labor to unionize the steel industry. Had

Midvale escaped the strike unscathed, the true value of employee representation, Dickson believed, would have been proved. As matters worked out, the test was not wholly fair to the ERP. The size and importance of the Cambria works at Johnstown were too great for the organizers to ignore. A strong campaign to unionize the plant was inevitable. At the same time, Midvale's top managers at Johnstown, though prepared to use any amount of force necessary to keep the union out, placed no faith whatever in the employee representation plan. A showdown was imminent.

CHAPTER

6

Test, Failure, and Collapse

THEY RODE in special parlor cars from Nicetown, Coatesville, and Johnstown to Atlantic City, accompanied by the managers of their respective mills. For two days they mixed business and pleasure while living at the luxurious Traymore Hotel, all at company expense. Meanwhile they received their usual daily wages, just as if they had remained on the job. Some of the men even brought along their wives and children as if for a holiday.[1] Being a shop committeeman under the Midvale Employee Representation Plan was not without its benefits.

The "most important accomplishment" of the Atlantic City convention, at least so far as Dickson and other company spokesmen were concerned, was the adoption of a resolution on how best to cope with the recent rapid rise in the cost of living. Proceedings of the Midvale ERP rarely made news except when local papers ran listings of recently elected plant committeemen or brief reports of what happened at quarterly sessions of the representatives in Philadelphia. The resolution adopted unanimously by the ninety-three shop committeemen in late August 1919—just a month before the great steel strike—was the exception.

"Midvale Steel Men Flay Wages Demands," declared the *Wall Street Journal*. "Representatives at Atlantic City gathering call wage extortioners profiteers." The *Washington Post* ran a similar account under the heading, "Labor's Demands Unwise, Claim Midvale Workers." "Steel Workmen Oppose Increase Pay Demand," reported the *Philadelphia Inquirer*. The *Pittsburgh Post* headlined its article, "Labor's Demands Uneconomic." [2]

The resolution noted that the "persistent and increasing demand" of all kinds of workmen for shorter hours and higher wages "to meet the high cost of living," was "uneconomic and unwise and should not be encouraged." After all, the average price of every commodity was determined by the average compensation that workers received for an hour's work. The only "sure remedy for the high cost of living" was "increased production" and the "stabilization of prices in conformity with wages now being paid." Workmen could best help by "diligent, efficient and conscientious labor, . . . thrift and the avoidance of waste and extravagance." [3]

Although the resolution closely paralleled Dickson's own views on the subject, he insisted that company officials, including himself, had had nothing to do with its formulation. Representatives from Johnstown, on their own, had discussed the problem and drafted the resolution prior to leaving for Atlantic City. At the conference the assembled commiteemen unanimously adopted it and ordered copies sent to the president of the United States, to senators and congressmen from Pennsylvania, and to state and municipal authorities. While Dickson may have had no part in shaping the resolution, he was responsible for the news release about it that resulted in stories in many leading newspapers following the convention. In his statement Dickson attempted to vest the resolution with as much significance as possible. Because the committeemen were chosen "at regularly held elections," he pointed out, their views were "therefore characteristic and representative of the thoughts and ideas" of Midvale's thirty-thousand employees.[4]

As a matter of fact, repercussions at Johnstown following publication of the resolution suggest that many of the workers there did not share the views of their representatives. According to the journal of the Amalgamated Iron & Steel Workers, employees at Cambria had been warned for a month or more that anyone who failed to report for work on Labor Day need not come in the next morning. In spite of the threat, large numbers of workers defied the company and took the day off—so many, in fact, that for the first time in its history the plant closed down for Labor Day. At the parade that day steelworkers carried banners, "We Are the REAL Representatives of the Cambria Steel Company," "We Are For Shorter Hours and More Pay," and "The Collective Bargaining Association [ERP] Must Go." According to the *Amalgamated Journal*, the Atlantic City resolution spelled the "suicide" of the company union. Milking as much from the incident as possible, the *Journal* charged that the delegates had all been handpicked by the company and made much of the fact that transportation to and from Atlantic City and all hotel expenses had been paid by the company. "To prove what might be accomplished by the company plan of collective bargaining," the *Journal* added gratuitously, "it is said that each delegate so inclined was provided with congenial feminine companionship. As might be expected, the company realized at once on its investment" by adoption of the resolution.[5]

It is quite possible that the resolution originated among the worker representatives. In addition to its more widely quoted portions, the resolution called for control of "private monopolies," restrictions on profits to "a rate that shall be fair to the consumer," a halt to the exportation of food and clothing from the United States, and the seeking out and placing on the open market of "all stores of hoarded supplies." Public officials were urged to exert every power to "bring about normal conditions, with special privileges to none but justice to all, and sure and swift retribution for those who may attempt to profiteer in the necessities of life." [6] Those portions of the resolution received relatively little attention from the press.

The most significant consequence of the resolution, however, was that it apparently disgusted and angered many highly skilled and highly paid workmen who previously had held themselves aloof from the organizing campaign. In effect, the resolution drove those men into the union movement. Dickson saw the resolution as a vindication of his faith in the essential community of interest between workers and their employers. Many, if not most, workmen saw it as manipulation by the company.

* * *

By the time that the National War Labor Board conferred its blessing on Midvale's scheme of collective bargaining in March 1919, the forces that eventually destroyed the plan were already at work. The end of the war four months earlier had removed the government's excuse for "meddling" in employer-employee relations. Not only were steel executives quick to question the right of the NWLB to push for collective bargaining, but the agency itself began to break up.[7] The armistice had also terminated the wartime pledges against strikes and lockouts. Even before the war ended, the American Federation of Labor had created the National Committee to Organize the Iron and Steel Workers and begun a struggle to unionize the industry. Steel companies, meanwhile, set about to smash such union sentiment as had sprung up and to maintain the open shop everywhere.

The most devastating result of peace for collective bargaining at Midvale was the impact on company profits, which fell sharply after 1919 and turned to losses in 1921 and 1922. Corey, Dinkey, and others had accepted employee representation under pressure. As profits failed, they quickly lost patience both with the scheme and its chief advocate, Dickson. As he had at U.S. Steel, Dickson urged that the profits crisis be met by the prescriptions of classical economics: cut prices so as to draw more business. Corey insisted that labor would not accept the necessary wage reductions until it had been "disciplined by shutting down mills." Dickson protested. Closing down would spell ruin to the towns of Coatesville and Johnstown, which depended on Midvale operations. At least the company should consult with its employees first.[8]

As a matter of fact, Dickson had already addressed a meeting of Coatesville businessmen and steelworkers. Demand for steel had fallen, he told his audience. More efficient mills across the country, including Midvale's Cambria Works at Johnstown, could turn out steel plate more cheaply than the Coatesville mills. To avoid a shutdown, Coatesville would have to reduce both its prices and its labor costs. Local landlords and merchants would have to cut rents and reduce the price of merchandise and groceries since the steelworkers could and should not carry the burden alone. "I succeeded in convincing the workmen and townspeople that I was sincere in my utterances," Dickson afterwards wrote. Adjustments in wage rates were accepted and the mills continued to operate "on a limited scale." [9] Although no works closed down completely, layoffs continued in all Midvale plants during the first half of 1919.

Meanwhile, in October 1918, the National Committee to Organize the Iron and Steel Workers had begun its drive at Johnstown. According to William Z. Foster, secretary-treasurer of the committee, the project began shortly after Midvale announced that it was establishing its employee representation plan. Workers, sensing that "something was wrong," decided that they preferred a "bona fide" labor organization or none at all. They invited Foster to start work among them.[10] Membership in the union grew. By March 1919, committee members claimed that between 40 and 50 percent of Cambria's employees had joined. Midvale officials insisted that the figure was closer to 10 percent. From the beginning, the organizers were concerned about the ERP.[11] Although they regarded it as a device for keeping out unions, they decided "to give it a fair trial by campaigning for and electing as representatives, union members who would demand the eight-hour day." At the regular annual election of representatives on January 13, a number of union candidates were successful.[12]

Even before the election, Corey was "very much excited" over developments at Johnstown and had demanded that Dinkey and Slick take action at once to remedy the situation. Slick declared war on the union by a wholesale firing of union members without regard for years of service or family need. Particularly singled out for discharge were union officials and union members who had been elected as employee representatives. Slick also had as many as a thousand loyal employees sworn in as deputy sheriffs and laid in a store of rifles to protect the company's property in the event of trouble. The National Committee complained to the Labor Department about Cambria's discrimination against union members, and on March 15 an official of that department brought the matter to Dickson's attention. At about the same time, O'Brien's published findings, which confirmed many of the complaints at Cambria, came to Dickson's hands.[13]

Following hurried conferences with Corey, T. L. Chadbourne, and other company officials in New York, Dickson went to Johnstown to investigate in person. "Slick is making a great mess of the whole situation including the collective bargaining scheme," Dickson recorded in his diary. When he discovered that Slick was preparing for a confrontation with the men not unlike that at Homestead in 1892, Dickson sent a messenger to Corey, recommending strongly that no attempt to use force be made in the event of a strike but that the company "rely entirely on public authorities to maintain order and protect property." Apparently with Corey's approval, Dickson discharged the deputy sheriffs that Slick had employed and disposed of the arsenal.

Further investigation, Dickson noted cryptically, revealed "records of men paid off, very loosely kept." Proceeding to Philadelphia, Dickson met with company officers and it was agreed that Slick should be asked to resign "for the good of the service." [14] At Dickson's urging, Alfred ("Fred") A. Corey, Jr., superintendent of U.S. Steel's Homestead works and the younger brother of Midvale's chairman, William E. Corey, replaced Slick as president at the Cambria mills. Dickson also took steps against possible troubles at Johnstown. Writing to county officials and conferring in person with the governor and attorney general of Pennsylvania, he explained that in the event of a strike Midvale would simply close down. The company would hire no guards other than those already employed to watch for fires and to prevent theft. Instead, Midvale would expect the state and local governments to protect its property and would "hold them to strict account for any damage." [15]

On April 7, only nine days after taking office, Fred Corey wrote a conciliatory letter to union officials in Johnstown. He would "gladly deal with employees" through the existing employee representation plan, he said, "or through any other accredited committee elected by the men in any way that is agreeable to them from among their own number to represent their interests in any matters they desire to take up with the management." Although he was later accused of having promised to work with the unions per se, his letter had gone on to say that he would "not deal with any outside organization or committees on questions involving the relations between the Company and its employees."

The new president probably was acting on his understanding of Midvale's labor policies as Dickson had explained them to him. Slick, he knew, had been discharged, among other reasons, for not working with the employee representatives. Dickson, if he was as expansive with Corey as he had been a month earlier with O'Brien, probably added that he saw nothing wrong with the plan evolving into a regular union. Certainly Dickson would have had no objection to union members serving as employee representatives so long as they were duly elected under his plan. Corey had gone too far, however, when he said that he would work with men elected other than through the plan. And so on May 7, in another letter to union officials, he explained that he would bargain collectively only with representatives chosen under the Midvale plan.[16] Despite continuing agitation by union organizers at Johnstown, the Cambria mills remained relatively quiet until the great strike in September.

Outwardly, Dickson appeared optimistic as to the future of the Midvale employee representation plan. In January 1919 he discussed collective bargaining at a meeting of the American Society of Mechanical Engineers and in June published an article, "Getting Our Men to Help Us Manage," in *System* magazine. Increasingly, however, Dickson came under fire from his associates for his labor views. On one occasion in March he told them that they were "all Bourbon reactionaries and out of touch with the spirit of the times." Their chiding, nonetheless, distressed him. "As long as I feel that I am really accomplishing something, I propose to hold my position, but at times I feel very much isolated from my associates and very much alone."[17]

The company meanwhile continued to make major decisions and to launch new programs after little more than token discussions with the worker representatives. For example, on May 7, 1919, only three days before the regular quarterly session with the representatives, the company adopted the pension plan that had been discussed only briefly with the representatives in February. Company officials, on their own, fashioned two other programs, a uniform mutual benefit plan that would cover all Midvale employees and replace the existing separate plans in the various Midvale properties, and a program for improving employee housing. On June 4, less than a month after the Spring session, the board of directors appropriated $10.5 million for reconstruction of plants, chiefly at Johnstown. Clearly the company regarded this matter as one strictly of concern to management.[18]

Although the National Committee's organizing activities were in full swing at the time of the ERP meeting in May, and the demand for an industry-wide steel strike was nearing fever pitch by late August, the calm of the Midvale quarterly sessions remained unbroken. The spring meeting was held at the Bellevue-Stratford in Philadelphia on May 10, the Cambria representatives coming over from Johnstown in Pullman cars the night before. "Rainy weather prevented sightseeing in Philadelphia," the *Johnstown Tribune* reported, "but the committeemen enjoyed the social diversions despite the unfavorable weather conditions." At the business meeting the proceedings involved more formal presentations than they did give-and-take discussions. "Five or more committeemen from each plant delivered addresses," and company officials introduced new programs. As the *Tribune* noted, it was evident the company had "already arrived at a solution" to the housing problem: Fred Corey announced that loans would be available to employees who wanted to build homes. The company also unveiled its new mutual benefit plan.[19] The account in *Iron Age*, probably based on information supplied by Dickson, attributed expansion of Midvale's housing program to a request from worker representatives. The *Tribune*'s account of Corey's presentation and the fact that the Midvale board of directors adopted the program less than a month later, suggests that the plan had been far advanced by management prior to the meeting in May.[20]

According to the *Johnstown Tribune*, worker representatives from the Cambria mills made one important contribution. They proposed that the August conference be held on a weekend at Atlantic City.[21] Even this alleged instance of worker initiative did not ring true. The company, after all, would be paying the bills. It seems less than likely that a shop committeeman would propose meeting in Atlantic City unless a company official at least intimated a willingness to meet in a resort setting. Whoever should be given credit, the conference at the ocean-front hotel and its celebrated resolution against higher pay and shorter hours roiled rather than calmed Cambria steelworkers and helped the cause of the union organizers.

As the great strike drew near, Dickson found his colleagues to be thorough reactionaries "who, if they dared, would revert to the high handed autocratic labor policy with which they were identified in the Carnegie Steel Co." Dickson called on them to honor and make use of the employee representation plan. "We have a definite, clearcut, agreement with our employees to confer with them through their elected representatives on 'all matters pertaining to conditions of employment, the adjustment of differences and all other matters affecting the relations of the employees to the company.' We must meet this obligation just as scrupulously as we would that of a financial one at a bank." He recommended an immediate conference at which the company would offer to institute the eight-hour shift system without further delay. The grant would be conditional, however, on a "substantial increase in efficiency per man." If efficiency did not follow, the plants could not be operated at a profit and the alternatives left to the company would be to reduce wage-rates or shut down. Again, Dickson's colleagues did not agree. Corey objected to any increase whatever in payroll costs. The "bourbon-reactionary attitude . . . evident throughout the discussion," Dickson observed, was "tempered only by a growing appreciation of the inadequacy of old methods in dealing with this great problem of labor and social unrest." [22]

Once the industry-wide strike began on September 22, Dickson sent recommendations to T. L. Chadbourne, Midvale's counsel, who was to participate in the Industrial Conference called by President Wilson for October 6. Dickson suggested that Chadbourne introduce a resolution recognizing collective bargaining "as a legitimate outgrowth of American ideals," calling for establishment, by law if necessary, of the eight-hour work shift in continuous operation industries and abolition of the seven-day workweek. Regardless of the strike's outcome, Dickson predicted, this program would eventually have to be adopted by the industry.

Moreover, by taking this stand, Dickson pointed out, Chadbourne would demonstrate his "friendly attitude towards labor," and would then be in a position to take a strong stand against recent activities of the A. F. of L., William Z. Foster ("the notorious I.W.W.") and his associates. The greatest issue facing the nation was whether the government was a real government or a "mere shell with the real authority lodged in the hands of Samuel Gompers and his revolutionary associates." In this struggle, Dickson declared, he stood shoulder to shoulder with Judge Gary. Once the strike was over and won, however, he proposed "to renew on every proper occasion," his "warfare against the seven day week, the twelve hour day, and the arbitrary handling of labor." [23]

Dickson believed that his policies were at least partly vindicated during the strike. At Nicetown, operations continued without interruption. At Coatesville, 30 percent of the employees went on strike for about two weeks. Although there was trouble at Johnstown—self-appointed "citizens groups," for instance, hampered the work of outside organizers—Dickson's no-confrontation policy was adhered to and bloodshed avoided. Fred Corey, whose allegedly enlightened views on labor had won him Dickson's support, now advocated abolition of the ERP because of the strike. To Dickson's surprise, Dinkey disagreed. Instead, the company notified its men that the mills would suspend operations "until employees in sufficient number expressed the desire that work be resumed." The "main issue" of the strike was whether the A. F. of L. would be recognized "as the representatives of the workmen of this Company and the controlling factor in the conduct of all matters of mutual interest to the workmen and the company." Midvale already recognized the right of workers to bargain collectively and currently was operating under a plan that it intended to use in the future. The Cambria mills remained closed until mid-November.[24] Although clearly defeated, the national leadership did not declare the strike over until January.

Dickson regarded the strike as a test of his collective bargaining plan and went to some pains afterward to explain why the ERP had not prevented difficulties at Johnstown. Writing in December 1919, he declared that in part the failure was the fault of managers at the Cambria mills who were "not in sympathy with the plan of representation, and by arbitrary and autocratic handling of delicate situations," had turned the workmen against the plan. The other factor was the realization by Foster and the A. F. of L. that "the successful operation of so-called Company Unions would be a death blow to the arbitrary power which they were endeavoring to exercise." For six months, Foster and his associates had concentrated on organizing the Cambria mills. Abolition of company unions was one of the demands of the National Committee.

During the strike, steel officials repeatedly had blamed the troubles on foreign workers whose ways were alien to the American system; Dickson had not. By February 1920, however, he too succumbed and added "the foreign element" as a third factor in the strike at Johnstown. Aliens, he declared, mostly from the "Slavonic nations," with "the Soviet idea . . . firmly implanted in their minds," saw the strike as a means for eliminating the bosses. Although a minority, this group, by threats of violence and the usual tactics of terrorism, were able to wield an influence very much out of proportion to their numbers."[25]

Dickson's lapse was not permanent. More in line with his usual thinking about foreign-born laborers was his response to the statement that workers could not be trusted because so many were "foreigners . . . not controlled by American ideals or sentiments"; a class "most susceptible to such non-working Bolshevik agitators" as "Lenine [*sic*], Trotsky, Debs, Hayward [Haywood], et al, et al."

I have lived for thirty six years in and about Pittsburgh [Dickson wrote]. I worked and ate and slept with these foreigners during the most impressionable years of my life. There are few places on earth where more material wealth has been produced as in that section. How much of that wealth was returned to the community in the elevation of the workingmen's standard of living? . . . [M]y duties took me frequently through the mill towns of the three valleys and I can still recall my disgust at the squalid living conditions which were so marked and characteristic of these homes. . . . [I]f these men were or are 'not controlled by American ideals,' whose fault is it? Who brought most of them to this country? Who maintained working conditions which tended to brutalize the body and soul? Answer—Carnegie Steel Company and U.S. Steel Corporation.[26]

* * *

If ever there was an opportunity for employee representation or company unionism to take root and become effective in the steel industry, it was in the wake of the 1919 strike. At least this seems to have been true at the Cambria works in Johnstown. Interviews conducted there in 1920 by the Interchurch World Movement revealed a strong distrust of unions among workers and deep seated prejudices against "foreign" and black coworkers. Enmity among various groups of workers in the same plant, of course, spelled trouble for any effective union organization.

Many of the men who had been on strike claimed afterward not to have believed in unionism. "The trouble with the union is . . . that it does not consider the company," one workman declared. "A strike is futile," another observed, "because the company just won't give in to the men." One embittered worker was quoted as saying that if anyone "approaches him with the subject of the strike he would strike him in the eye." Union officials came in for particularly harsh criticism: "the organizers get all the money and don't do any work for it." One worker believed that Johnstown union officials had "absconded with $65,000 belonging to the steel mill." [27]

Not all "American" working people disliked foreign-born workers. One worker's wife expressed "great admiration for the foreigners because they stuck like leaches" during the strike. Most other opinions were hostile. "The Foreigners were going to run the mills themselves," according to a rolling mill employee. "They had their own superintendents and other officers all picked out." A rolling mill finisher observed that foreigners were "good workers, but misguided." Several of the people interviewed reported that foreign workmen were leaving Johnstown, some to visit relatives in Europe now that the war was over and some to escape the rigors of the new prohibition amendment.[28] Blacks, hired as replacements for the foreigners, drew uniformly unfavorable comments: "they are a worthless lot, . . . they are absolutely no good as workers," they "don't even earn their transportation," they "are of no account at all as workers, . . . they don't even mind if you dock them," "they simply won't work." [29]

It was in this climate that the first poststrike election for plant committeemen was held in January 1920. According to the *Johnstown Tribune*, delegates were elected in "spirited contests" with an "unusually large vote." Chiefly new men were elected, only four of forty-one having previously served as representatives. "For the first time since the organization of the Collective Bargaining Association," the *Tribune* continued, "the general mass of the Cambria employees have shown a real interest in the organization and have thrown aside all doubts as to the possibilities of the organization." [30] While the article may have been inspired by Cambria officials (during the strike the *Tribune* came close to being a company organ) and not representative of the views of workers, the top managers at Cambria only tolerated, did not encourage, employee representation. It is possible that workers decided to give the plan a chance, seeing it as better than no organization at all.

In August, Cambria officials and all plant committeemen again joined their counterparts from the other Midvale plants for a convention at Atlantic City. The *Tribune*'s account of the departure from Johnstown revealed why it was difficult to take employee representation seriously. "It was intimated that the representatives would discuss the wage question, hours of labor and welfare measures at the convention. However, it was stated that there had not been prepared any special resolutions or a set program for the session. Many of the representatives consider the convention more as an outing this year than a meeting for any serious business matter." [31]

That employee representatives thought of the organization chiefly in terms of the small pleasures that it brought them, at company expense, was hardly surprising. Between the close of the 1919 steel strike and the demise of the Midvale Steel & Ordnance Company in 1923, Dickson alone fought to preserve the plan. By January 1920, Fred Corey had written off collective bargaining as a failure. The "natural tendency of the times," he believed, was against "unreasonable labor demands," and he favored a campaign by businessmen to "enlighten the public." With a little effort, "we could look forward with confidence towards the ultimate outcome of the Bolshevik or labor movement (which is one and the same thing)," he argued, and enjoy a "very lengthy period" of industrial and business peace.[32]

By belittling employee representation to his staff, Dickson warned,

Corey was "following in the footsteps of the unlamented Slick" and making a fair administration of the plan impossible. Dickson did not expect "any immediate and wonderful results" from collective bargaining, but neither did he think that the "hands on the clock of civilization" would be turned back "so as to perpetuate the autocratic system of handling labor." Workmen perhaps did not know how to use their new freedom wisely, he conceded, but it was Corey's duty "as their natural leader to show them how, and to do this *with the utmost patience.*" Dickson predicted that Corey would find educating superintendents, foremen, and gang leaders more difficult than training workmen in the "American method of handling labor." The remedy that he recommended was extreme: "just as soon as I saw that an official was knowingly bucking the plan, I would cut off his official head so quickly that he would be on his way before he knew what struck him." [33]

As the February meeting with plant representatives drew near, Dickson told Fred Corey that he had no "new message" to deliver. What did Corey think of inviting the men to cooperate in publishing a monthly magazine for distribution to all employees? "We hitched our industrial wagon to a very live team when we adopted the Plan of Representation," Dickson declared, and "we must either guide the team with skill, or be pulled over the dashboard by it." Midvale had already "failed to measure up" to its opportunities by cutting the number of company-wide meetings with representatives from four to two a year. "The men are eager for elbow touch, and we ought to be sure enough of ourselves, and resourceful enough, to respond to this need." By April, Dickson professed agreement with Corey that "propaganda of a popular kind" was needed "to counteract the un-American ideas . . . being disseminated by half-baked agitators." The best way to contain "a prairie fire," however, was backfiring. By making its employee representation plan work well, Midvale would be contributing handsomely to the campaign "against the dangerous elements in society." Again he urged a company magazine for educating employees. "The great majority of our men are real Americans, who, while they will resent any attempts at autocratic control, are eager for real leadership." [34]

"The only thing wrong with your suggested plan," Corey replied, "is that it won't work." Times were going to be "rough" until "we get down to first principles again and get to work," he declared. The "only real convincing message to persuade men to work," he insisted, was to have them find it "necessary to work in order to eat, much less own and ride in automobiles and drink liquor and wear diamonds." Corey, in negotiating for his position at Cambria, had held out for a three-year contract, guaranteeing him $100,000 per year in combined salary, bonuses, and stock. The average annual income of the men for whom he recommended austerity was at an all-time high in 1920—$2,161. The average number of men employed, however, had fallen nearly 20 percent since 1919.[35]

After 1920, Midvale's earnings began to run far behind expenditures, and the austerity called for by Corey became fact. Employment fell nearly 50 percent between 1920 and 1921, and the average annual wage per employee dropped to just slightly over fifteen hundred dollars. Meeting with the plant conference committee at Nicetown in July 1921, Dickson found the representatives "very much disgruntled," and, in his opinion, "justly so." The company, arbitrarily and in an untactful manner, had reduced wages and cancelled the yearly conference with employee representatives at Atlantic City. "If the Plan of Representation fails," Dickson concluded, "it will be because of the lack of ability of the management to meet the men half way." [36]

By August 1922, Fred Corey and his allies in the company were once more trying to cancel ERP meetings altogether. Corey sent Dinkey a photograph of some Johnstown workers, labeled on the back, "Slickville Hun Invaders, who must receive at least $7.50 per day to maintain an American Standard of Living. Try collective bargaining with the son of a bitch in the noose." A noose was sketched in ink around the neck of one of the men. Dickson, to whom Dinkey passed the picture, noted, "This is a shining example of the cynical attitude of my associates . . . toward our plan of Employee Representation. What results could be expected from such leadership?" [37]

Dickson's commitment to collective bargaining under the ERP and his sincerity in dealing squarely with worker representatives was beyond question. His associates in the industry rated him somewhere between "impractical idealist" and "dangerous radical" on the labor question. One asked if his middle initial, "B," stood for "Bolshevik." The leadership that Dickson provided at meetings with worker representatives, however, and his views on the uses that might be made of the ERP show how far short his idealism fell from meeting the needs and interests of the workers.

Among the schemes that Dickson favored for resolving the labor problem were stock ownership and profit sharing by employees. Neither idea was new and neither had proved particularly effective in the past. Under the Midvale stock purchase plan, for example, after several years of operation in a period of unusual prosperity, only 365 (or slightly over 1 percent) of the firm's employees actually owned shares.[38] Dickson's interest in profit sharing dated from 1919. Pressing for that program at Midvale in the early twenties, when there were no profits to share, must have seemed the height of impracticality to his bemused colleagues.[39]

Still, it was employee representation that seemed to hold the most promise from Dickson's point of view. Apparently assuming a complete community of interest among the employing and the employed, Dickson believed that Midvale's Employee Representation Plan could be used for politicial as well as industrial ends. By "acquiring a real legitimate leadership over our 40,000 [*sic*] employees," he told William E. Corey in February 1922, "we could talk to the politicians with a force behind us which they would respect." Corey thought differently. The "only way to interest the politician," he observed, was "by buying them." [40]

Perhaps Dickson's most persuasive argument in favor of employee representation, at least from his associate's point of view, was that it tended to forestall regular unions. Dickson himself became increasingly hostile to unions during the early 1920s. His onetime respect for Gompers and the A. F. of L. faded rapidly during the steel strike. Finding himself unexpectedly on the same public platform with Gompers in late 1920, Dickson condemned the labor policies of both the old Carnegie Company and its successor, U.S. Steel, as feudalistic. Not wishing to be "misunderstood," however, Dickson turned to Gompers and said, "I desire to make it clear that in the recent strike the course taken by the Steel Corporation had my hearty approval." [41]

The rash of strikes in those years, and especially the steel strike of 1919 and the coal and railroad strikes of 1922, all deepened Dickson's distrust of unions. "We have already lost about $1,000,000.00 since April 1st by the coal strike," he complained. "If more money is to be expended I prefer to expend it in fighting the autocratic Miners Union." [42] That same month, July 1922, Dickson called for the decentralization of unions in a letter to the *New York Times*, copies of which he sent to President Harding and to selected senators and congressmen. Public opinion, Dickson declared, had already "rendered an adverse verdict against unrestricted combinations of capital." In response, Congress had restrained such combinations by the Sherman and Clayton Antitrust Acts. Without stopping to discuss the effectiveness of those measures against combinations of capital, Dickson called for similar laws against the "arbitrary and selfish use of power" by unions, especially where the public welfare was involved. Under existing law, he pointed out, it was illegal for two companies to fix the price of the products they sold. At the same time, "an irresponsible organization of miners, the United Mine Workers of America," which was "not even incorporated," had the "legal right to organize the employees of all the coal mining companies in the country," and thus dominate the production of this essential commodity. While legislation "should jealously guard the right of workmen to organize for common defense," Dickson believed it should also bring unions under "the same Federal laws which safeguard the public from combinations of capital." [43]

It was at meetings with the Midvale employee representatives, however, that Dickson revealed the great gulf that had grown between his outlook and style of living and that of the workers from whom he boasted having sprung. "The strenuous man is the happy man," he told the plant committeemen assembled at Atlantic City in August 1920. If he could, he would add to the Beatitudes "Blessed is the man who has found his work." Dickson went on to warn the workmen that life consisted of much more than the acquisition of material goods. To possess things, he cautioned, often meant being possessed by things. Finding this true in his own life, he confided to his audience, he had built "a rough shack" in the woods at his New Hampshire farm. "It contains only bare necessities in the way of furnishings," he said, "and I have told my wife that I will shoot anyone who dares to bring anything into that shack which has not been proved, beyond question, absolutely essential to my comfort or happiness." There, periodically, he retreated with his family to escape the choke of possessions. After the session the worker representatives returned to the mills, presumably to continue their quest for possessions, while Dickson went to his New Hampshire retreat to get away from the burdens imposed by responsibility and worldly goods.[44]

In most of his formal addresses to employee representatives, Dickson tried to educate the men on the economics of Midvale's operations. As he explained in an article in *System* magazine in June 1919,

> The typical workman now has only the haziest ideas as to the nature and function of the business he is in. Often he thinks of the resources as a kind of inexhaustible resevoir, from which the employer can raise wages as much as he likes by simply opening the spigot wider. Look into almost any workman's thought on the subject, and you will find something like that. Small wonder the demands of labor are often unreasonable.
>
> The workman needs to know more about the actual problems of management. He needs to learn something about overhead, about marketing difficulties, about the dependence of production and production conditions upon marketing and the dependence of marketing upon economic production, about the numberless hazards and chances of loss which his employer must face. He needs, in short, more of the manager's point of view. And the obvious way to give it to him is to let him have some part in handling management problems.

In August 1920, he lectured the employee representatives on the roles of stockholders, managers, and workers, and the sources and distribution of the company's income. At the February 1921 session, Dickson explained why Midvale had to reduce prices (and hence wages) drastically in order to attract new business. One year later, using the company's balance sheet as his text, he explained "What Happened to Midvale in 1921?" For the last session that he attended, Dickson wrote a short play, "The Three Must-Get-Theirs," which worker representatives performed. The message of the sketch was that capital, management, and labor had to work together, not only for their mutual welfare but for the public good as well.[45]

In his diary, Dickson often remarked on the successes of these sessions. It is unclear, however, what the meetings produced beyond the "elbow touch" that he believed was needed by both workmen and managers, a chance for the company to explain its position to the men, and an opportunity for workers to preview some of the company's worker welfare schemes. Such successes as the Midvale ERP produced came not from the general meetings but at the individual plant level. During the first nine months of operation at Johnstown, for example, the plan adjusted "hundreds" of disputes over working conditions and wages, "some of them of many years' standing." Toilets, lavatories, and baths were installed for the convenience of workers. A home building loan program, financed by the company, was instituted "at the request of the employees' representatives." Baseball and football teams were started, and the company opened a free bathing beach to employees and townspeople. Even the company's reservoirs were opened to employees for fishing. The mere existence of grievance machinery seemed to reduce instances of abuse by foremen and department superintendents and to give workers some sense that they would be treated fairly.[46]

Except for the resolution of grievances, however, these "reforms" were common features of welfare capitalism and were in no way particularly dependent on worker representation. At no point did the Midvale plan show any signs of evolving into a system of full and equal partnership between management and labor. It was not that the plan was undemocratic in form. The deficiency lay in the lack of substance behind the form. Workers, for example, had not conceived the plan—they, or at least their representatives, had merely ratified the scheme that Dickson had adapted from the Colorado Fuel & Iron Company's ERP. Meetings were held in places unfamiliar to workers, where they were surrounded by unaccustomed luxury and comfort at company expense. No leadership evolved among the worker representatives during the five years that the plan was in effect.

With the possible exceptions of the home-loan program and the celebrated 1919 resolution against higher wages and shorter hours (both allegedly called for by worker representatives) no program was ever initiated from below. No reported conflict arose at any of the sessions between the managers and shop committeemen. That such elementary concerns of workers as wages, hours, seniority, and protection against layoffs were never discussed indicated that the representatives regarded themselves as powerless in these critical areas, or feared they might antagonize company officials. In an era of falling wages, extensive layoffs, and continued operation of the twelve-hour shift and seven-day workweek, Midvale's ERP failed to represent the interests of the rank and file.

Dickson always expressed hope that the plan would produce true labor-management teamwork. Either he had no skill in drawing out the representatives and encouraging them to function as spokesmen for labor or he was blind enough to believe that his homilies answered any questions they might have. That he regarded the sessions as "successful" suggests that, whatever he professed, he was content to soothe the workers and keep them passive, not really to share responsibility with them in the quest for solutions to common problems.

Both Dickson's opposition to autocratic labor policies by management and his abhorrence of an adversary relationship between worker organizations and management were understandable and even commendable. His dream that capital and labor could reason together and cooperate in the production process, however, was doomed to remain a dream unless labor could feel free to participate as an equal without fear. Given time and sincere intent and active encouragement by an enlightened management, it is possible that Dickson's plan might have developed into a true partnership of workers and employers. The history of company unionism, however, suggests otherwise. In the end the Midvale plan remained the aspiration of a single man. The workers went along for the free ride, having little to lose. Dickson's colleagues and superiors at best tolerated the pet whim of their idealistic associate only as long and only to the extent that it did not add much to costs or threaten their absolute control of the company.

* * *

When they established the firm in 1915, Midvale's promoters insisted that the company would produce a wide variety of steel products, not just war materiel. Despite a slackened demand for steel after the armistice, Midvale pushed ahead with reconstruction projects at its Cambria mills, acquired new ore properties, and talked of building a new plant on the Atlantic seaboard. The loss of military and naval orders, however, immediately played havoc with the company's balance sheet. Earnings for the last full quarter of war production (July, August, and September 1918) amounted to $3.89 per share. The final quarter of 1918 brought in only $2.51 per share. In all of 1919, Midvale shares earned but $5.30 as compared with $17.79 in 1917 and $14.60 in 1918. Operations in 1921 and 1922 lost money. Dividends followed a parallel course. Earning over 35 percent on its investment in 1917 and 29 percent in 1918, Midvale paid a regular dividend of 12 percent. In May 1919, it cut the rate to 8 percent, in January 1921 to 4 percent, and after April 1921 it paid no dividends at all.[47]

The company's dependence on war contracts made it particularly vulnerable, but it was not alone in its affliction. According to *Iron Age*, the industry at large operated at only 38 percent of capacity in 1921.[48] As Dickson analyzed the situation, the industry's problems were aggravated by U.S. Steel's dominant position and by Judge Gary's unreasoning and persistent commitment to price and wage-rate stability without regard for changing economic conditions. By the end of 1920, demand for steel had fallen to half of normal—or to just about the level of U.S. Steel's capacity. Since the corporation could produce more cheaply than its competitors, Dickson reasoned, independents would sell little until U.S. Steel's sales neared its productive capacity.[49] Bearing out Dickson's contentions, early in February 1921, the corporation's operations were "still at an 80 to 90 per cent rate," while the independents were running at between 20 and 35 percent of capacity.[50]

Dickson attributed U.S. Steel's advantage to several factors. Its scale of operation, for example, made profitable the conversion of wastes into saleable by-products such as fertilizer and cement. The corporation owned or controlled a fleet of forty ships engaged in overseas trade. Its great diversity of product increased sales because, all else being equal, customers preferred doing business with a firm that could supply all of their needs. The leading factor, however, was U.S. Steel's ownership of the two railroads linking the Mesabi Range with Lake Superior, a fleet of ore boats on the Great Lakes and the railroad that connected Lake Erie with the Pittsburgh steel district. The corporation not only hauled its own ore at cost, it realized a profit of $1.20 per ton of pig iron for transporting raw materials for its competitors.[51]

When independent steel executives tried to stir interest in a federal law that would have forced U.S. Steel to give up its railroad properties, Dickson went along.[52] At the same time, however, he resumed his attacks on Gary's policy of price and wage-rate stability. Dickson first called on his own company to reduce prices. The fortunes of Midvale were at low ebb, he pointed out, with only the Nicetown plant operating normally. All but two of fourteen blast furnaces stood idle, and the plate mills at Coatesville and several departments at Johnstown were shut down. Midvale's management, Dickson argued, was trustee not only for the firm's stockholders but also for employees and their families who depended on the company for their daily bread. The one way that the firm, as an independent producer, could attract enough orders to keep its plants open and providing a livelihood for its workmen and their families was by reducing prices drastically.

Since 85 percent of the cost of steel went for labor, Dickson noted, wages also would have to be lowered. The company could force reductions by keeping its plants closed until the men, out of "sheer necessity," accepted lower rates. A more reasonable course, however, would be to explain the situation frankly to the men and seek their cooperation. It would have to be pointed out to them that the choice was not between work at existing rates and work at reduced rates, but work at lower wages as against no work at all. "You do not pay your household bills with wage rates, but with actual wages received," Dickson proposed telling the men, "and if the mills remain closed, or operate only part time, there will be little or no actual wage payments." Because of its commitment to the ERP, Dickson insisted, Midvale had no alternative but to take up the matter with its men.[53] His associates did not agree that the workers should be consulted. "It all seems so crude and barbarous," Dickson wrote in his diary. "Why can we not get close enough to our workmen so that we can reduce wages to a minimum at such a time and so be able to run at least part time? We need education and vision at both ends of the scale and I am somewhat hopeless as to the upper end." [54]

Continuing depression helped persuade Midvale officials to lower both prices and wages. On Dickson's recommendation, they agreed to call a meeting of all interested steel producers for February 4, 1921. There Dickson announced Midvale's "drastic action" to officials of eleven independents, U.S. Steel being conspicuously absent. Although the causes of the slump were "very complex," Dickson conceded, the "psychology of the situation" was that "no buying of any consequence" would begin until customers were convinced that the falling market had reached bottom. Midvale neither asked nor expected the others to follow its lead, Dickson declared, but the longer the industry delayed cutting prices, the longer the depression would go on. "Our action today," Dickson recorded in his diary, "is in my opinion the most important trade action since the war." He was "proud of having been able to bring it about" in spite of Dinkey's opposition and the "lukewarmedness" of others.[55]

In reporting Midvale's new policy, the press predicted a general price war in steel. A few days later, however, Judge Gary, with his usual studied calm, told reporters that U.S. Steel had no intention of cutting either prices or wage-rates. Expressing surprise that the independents could afford to sell steel at prices below actual cost, he declared that any manufacturer who cut prices "must intend reducing wage rates" as well. That, Gary said, would shift the burden onto the workmen and would be "manifestly wrong and unfair." Every manufacturer "must know" that cutting prices would bring in no more business. Any advantage gained by one seller would be temporary at best and "the whole matter . . . evened up when his competitors cut prices." [56] Gary made no attempt to answer Dickson's point that high wage-rates meant nothing to the thousands of steelworkers who at the moment were laid off.

Most of the independents, following Midvale's lead, cut prices between 15 and 20 percent. The expected flow of new orders, however, did not materialize, and steel buying, nationwide, dropped to below one quarter of the industry's capacity. The cuts did bring limited benefits to the independents. Midvale's Cambria mills expanded operations from 40 to 60 percent of capacity, while independents generally increased their share of the falling demand from 30 percent in January 1921, to 35 percent by April 1, and to 40 percent by April 25. Gary met this threat to U.S. Steel's position just as he had during the similar struggle in 1909: he cut the corporation's prices on April 12 and wages on May 5. As the Judge predicted, no one gained much for very long, and the depressed state of the industry continued. When Gary announced his reduction in prices, the independents, who had been losing money at their even lower rates, raised prices to match those of U.S. Steel, and uniformity once more prevailed. The revolt of Midvale and the independents was over.[57]

A general improvement in the steel market in 1922 helped Midvale very little. Once the tide turned against the company, nothing seemed to go right. On April 6, 1921, the firm recorded two unfortunate "firsts": its first quarter of operation at a loss and its first passing of a dividend. Sickness and other personal problems soon began to plague high officials of the firm. On April 15, President Dinkey was temporarily felled by a stroke. About the same time the drinking habits of a top official at Johnstown gave rise to fear of a public scandal. The man came to a directors' meeting late and with "all the appearance of a 'hangover' from a last night's debauch." Three weeks before, "disgustingly drunk" at Johnstown, he had "made a spectacle of himself." The company, Dickson observed, "is suffering from his utter lack of fitness to occupy his position." Chairman Corey told the man either to "stop drinking or drink himself to death quickly." One year later Dickson was hospitalized with a bleeding ulcer. Some people, including Dickson himself, feared that he had "got his ticket." Rest and a restricted diet, however, corrected the problem without the surgery that had originally been prescribed.[58]

Meanwhile, events outside the business world began to afflict Midvale. When the Washington Conference began deliberating disarmament in November 1921, Dickson predicted that "if successful," it would "mark an epoch in history." The "first fruit" of the conference's naval limitations agreement hit Midvale's Nicetown plant in February 1922, when the government halted all work on armor and guns for the navy. The cancelation brought the company "face to face with a real dilemma—what to do with that plant." Some four hundred men were discharged, a quarter of whom were "super-annuated employees" previously kept on the payrolls "partly as a matter of charity." These men were not eligible for company pensions and once discharged were left without means of support. "This is a tragedy," Dickson declared. "The cost to the Company is about $1300.00 per week. Here is one of the weaknesses of our industrial system. Our stockholders are without dividends and cannot support these men. What is to be done?" [59]

Dickson eagerly sought ways to reverse the current running against Midvale. Since only about 5 percent of its business was in guns, he proposed dropping "Ordnance" from its name. Perhaps the company should try to break into the midwestern market, he suggested, by buying or constructing facilities in the Chicago or St. Louis districts. In January 1922 he recommended reviving Carnegie's old sliding-scale formula for wages, a scheme that would tie wage-rates automatically to the selling price of steel. As an economy measure, Dickson suggested closing down Midvale's New York and Philadelphia offices and concentrating the firm at Johnstown. "This means much to the members of our organization," he admitted, "but 'needs must when the devil drives.' " [60]

As the company's fortunes sank, Dinkey, the Coreys, Dickson, and other officials became edgy and increasingly gave way to mutual recrimination. "When I consider the material resources of Midvale," Dickson wrote in his diary, "I feel like borrowing money to invest in the stock, but when I consider the type of man who is President I want to sell short." Discussing the company's future with William E. Corey, Dickson pointed out that although the board of directors was composed of "big men," their influence on the firm's policies had been "practically nil." The board had "never suggested a wise move, nor prevented us from making a foolish one," he complained. With William Corey disinclined "to mix up in the turmoil of active management," and Dinkey neither comprehending the problems at hand nor "effectively" grappling with them, Fred Corey remained "as the only hope of the Company." If that was true, William Corey replied, "God help the Co.!" [61]

Midvale's sales manager, meanwhile, called for curtailment of selling (presumably because of low prices) and criticized the company for being "top-heavy, especially at Johnstown." Fred Corey, who was in charge at Johnstown, replied that the sales manager had "no constructive ability as a salesman," was "a mere office martinet," and should be "displaced by a man with more diplomacy and 'sauviter in modo.' " [62]

The gloom persisted, and in July 1922 William Corey placed responsibility for the condition of the company on the management's lack of foresight. Instead of rehabilitating its plant, the company should have conserved its cash resources. He blamed himself for not preventing the policy and Dickson for advocating the purchase of new ore properties. Dinkey, in turn, blamed his colleagues for his own failure to buy up scrap iron when the purchase could have saved the company nearly $1 million. He allegedly told Fred Corey that Dickson had opposed spending the money and later told Dickson that William Corey had not approved the purchase. "It is to laugh," Dickson observed, "a successful liar should have a good memory." [63] Clearly the strain was telling on them all. Whatever the truth of the various charges and countercharges, taken together they did not explain Midvale's continuing decline.

During the postwar slump, steelmen hatched innumerable schemes for restructuring the industry, at least that part not already controlled by U.S. Steel. "As a matter of fact," Dickson commented, "most of the independent steel companies existed only on the sufferance of the U.S. Steel Corporation." Midvale's position was particularly difficult. The Steel Trust's competition with Midvale went beyond the "natural desire to secure business." A "more or less personal animus" existed against the firm's officials, "all of whom had formerly been associated" with U.S. Steel.[64]

To escape its problems, Midvale sought to merge with other independents. After brief flirtations with Bethlehem in 1919 and 1920, Midvale officials pushed for a merger that would include Midvale, Lackawanna, Youngstown Sheet & Tube, Republic, Inland, the Steel & Tube Company of America, and Briar Hill Steel Company. The seven, as a single firm, would have had 20 percent of the industry's producing capacity as compared with U.S. Steel's 45 percent. In December 1921, Bethlehem, alone, stood second with 6.38 percent of the industry's capacity. Midvale was third with 5.76 percent.[65]

Conflicting ambitions among the steelmasters as to who would head the new combine, plus the fact that several of the firms were conducting simultaneous negotiations for other possible combinations, prevented agreement. When one steel executive was caught making offers to join two competing combines, a friend offered as an excuse that he was "not a gentleman and therefore did not realize that his action was questionable." [66] By the spring of 1922 the seven-company merger had given way to an agreement between Midvale and Republic to buy up Inland Steel and to form the North American Steel Company. Corey was to be chairman of the board, J. A. Topping of Republic, president, Dinkey, vice-president in charge of finances, and Dickson, vice-president in charge of sales.[67] Submitted to the Justice Department and to the Federal Trade Commission for approval, the plan was shot down by the FTC. Dickson was disappointed: "Am greatly surprised at this action on the part of my old friend Nelson Gaskell, Chairman of the Commission." [68]

Within the month, Corey arranged to sell Midvale's Coatesville and Cambria plants to Bethlehem. By November 1922, Dickson realized that his career in steel was ending. Neither his Midvale associates nor Bethlehem officials had talked to him about a place. "I know that my attitude against the seven day week and the twelve hour day has been strongly resented by CMS [President Schwab of Bethlehem]. . . . From the financial standpoint I have played my cards poorly. Here I am just entered my 58th year and poorer than at any time since I became a Carnegie Partner." [69] Dickson's poverty stemmed not just from the threatened loss of salary or even from the sharp decline in the value of his Midvale holdings. The previous May when he had entered the hospital, "not expecting to recover," he had distributed "most" of his estate among his five children and his brothers and sisters.[70]

Just prior to the takeover by Bethlehem, Dickson learned that he would probably be given $100,000 in cash but no position with the firm. As a matter of fact, he received $90,000 ("i.e. three years' salary") as "a gift to express the goodwill of Mr. Schwab . . . and the Bethlehem people generally." [71] On May 31, Dickson moved from the office at 14 Wall Street where he had spent seven and one-half years. They had been "in many ways the happiest of my life," he noted, "in spite of the fact that I have been thwarted in the full realization of my ideals by the unwillingness and inability of some of my associates to do their part. This is the end of another important epoch in my life. . . . I am again thrown on my own resources and very meager ones they are." [72]

Once more involuntarily adrift, Dickson began the painful adjustment to retirement on relatively limited funds. The scale of living to which he and his family had become accustomed made retrenchment difficult. Before retiring, Dickson laid plans for building another house on his New Hampshire farm. His idea was to make the place "self supporting if necessary," changing it from "a great liability, as it is at present, to an asset." [73] In April 1923, to his wife's sorrow, the mansion in Montclair was sold. "I have groaned under the burden of taxes and maintenance for so many years," Dickson observed, "that I welcomed the idea as a relief from a crushing burden. I am afraid that I have grown to have an intense dislike of the place." [74] Mamie Dickson gave a final tea for six hundred guests in May, then closed down the great house and moved with the family to New Hampshire. In December 1924 the Dicksons bought a much less expensive home in Montclair.[75]

For a while Dickson lost much of his great zest for life. "As the year closes," he wrote in his diary at the end of 1924, "I find myself strong physically but exceedingly restless mentally. I feel that I must have regular employment and am at a loss to know how to find it. . . . Here I am, in my sixtieth year, sound in body and I trust as mentally alert as ever I was, yet debarred from the active life to which I have been accustomed for the past forty-seven years." Dickson founded a choral society in Littleton—the "Highland Chorus"—that he directed, he began writing a column for the local newspaper, and he gave himself over to hard physical labor on his farm. But to no avail. "Nine more years due me according to the scriptures," he wrote on his sixty-first birthday. "Not specially interested." [76]

CHAPTER

7

What Might Have Been

AMID THE DISAPPOINTMENTS and frustrations of his "wasted life," as he sometimes termed it, Dickson lived to see himself repeatedly vindicated by later events. The battles he fought and lost—to abolish the seven-day workweek and the twelve-hour shift and to secure for workmen a voice in determining the conditions under which they labored—in the end were realized. Dickson's first belated victory came with the adoption of hours reform by the steel industry.

Dickson had resumed his attack on long hours shortly after becoming vice-president of Midvale. In December 1916 he addressed the American Association for Labor Legislation at Columbus, Ohio. To move to the eight-hour day would add considerably to the cost of steel and would be impractical so long as the supply of immigrant labor was curtailed by the war, he admitted. At the same time, he insisted that there could be "no supine flinching" from the issue. If the leading employers were not willing to cooperate in ending the twelve-hour shift, "then the power of government must be invoked to compel a readjustment to more humane conditions." [1]

In 1918 and 1919 Dickson saw what he thought were signs that victory was near. Under pressure from the government, Judge Gary in late September 1918 announced adoption of the "basic eight-hour day" throughout the steel industry. Although wartime shortages of labor precluded the men from working only eight hours, they were paid time-and-a-half for all hours over eight and at war's end, Dickson assumed, the eight-hour shift would become reality. As a tactic for warding off the 1919 steel strike, Midvale's board of directors authorized the officers "to establish the eight hour day when in their judgment the interests of the company required such action." "The world moves forward," Dickson commented in his diary, "and I can almost see the realization of my dream of the final extinction of that relic of barbarism in the steel business—the twelve-hour day." [2]

But the end was still not at hand. The bitterness engendered by the great strike did not incline employers to grant any concessions to the defeated workmen. Dickson persisted in his crusade. In an address before the American Society of Mechanical Engineers in November 1920, he criticized U.S. Steel for continuing the long shift. His comments on that occasion backfired. One Moses Cleveland addressed an open letter to him: "Dear Bill, . . . may we take this to mean that you intend to abolish the 13½ hour night shift at Johnstown, Pa., in the mills of the Midvale Steel Co., of which you are vice president?"

A friend enclosed the clipping and invited Dickson to react. Dickson was defensive. His witty reply carefully evaded the main issue.[3] Charles W. Baker, an intimate friend to whom Dickson showed the correspondence, was not impressed. "It seems to me that Mose 'has scored a hit on you.' . . . What I make out from your letter . . . is that you are trying to kick up a dense dust to cover an ignominious retreat. . . . Whatever your other faults, I never thought that you would accuse others of sins of which you yourself are guilty even in a higher degree."

As defense, Dickson offered a syllogism: "*Major premise*- Responsibility and power are one and inseparable; *Minor premise*- Your humble servant lacks power; *Conclusion*- He also lacks responsibility." Dickson suggested that if Baker read the formula over "carefully several times, and very slowly," he was sure "some light may begin to dawn on you." His attack on U.S. Steel, he pointed out, was meant to serve as an indirect prod to his colleagues at Midvale.

"I take it that you wish to plead non-responsibility for conditions at home," Baker replied. "This may be a good plea in fact but to an outsider . . . who does not know this but presumes the contrary from your exalted position, it looks bad. . . . A flank attack is often justifiable, but a direct assault is the more heroic and the one that arouses admiration." [4]

Dickson indeed was powerless to bring about reform at Midvale where President Dinkey persisted in upholding the long shift. Dinkey "reiterated his reactionary views," Dickson noted on one occasion, "stating among other things that the men were better off in the mills than they would be in their homes or elsewhere." Dickson told Dinkey and a friend with equally unenlightened views that "their brains belonged in a museum." [5]

At U.S. Steel, Judge Gary, too, kept up the fight to preserve the long workday. Criticism of the corporation's labor policies, he told stockholders in April 1921, "generally originated with" or was supported by "ill-advised or vicious-minded outsiders and not by the workmen themselves." The corporation had inherited both the twelve-hour shift and the seven-day workweek. "Perhaps they will never be entirely abolished," he added. "Possibly the workmen themselves, the employers, or the general public will never, as a whole, consent to the entire elimination of either proposition." Gary's statement sounded "like a ukase of a Russian Czar," Dickson commented. His further declaration that stockholders "properly may and ultimately will dictate . . . rates of compensation to employees, . . . terms and conditions of employment," sounded to Dickson "like a 'voice from the tomb' of our boasted liberty." [6]

Although the victory of the steel companies in the great strike of 1919 forestalled unionization of the industry for a decade and a half, it did not end the demand of reformers for redress of the grievances that had sparked the organizing drive. On July 28, 1920, a commission of inquiry of the Interchurch World Movement issued a report blaming the strike not on "ignorance" or "bolshevism" among the workers, as the companies charged, but on the "barbaric" twelve-hour shift, long turn and seven-day workweek, on low wages, on the lack of grievance machinery, and on the arbitrary and autocratic domination of the workers' lives by a handful of New York financiers. The forty-one pages of the report devoted to "The Twelve-Hour Day" exposed the error of the widely held notion that the long workday had disappeared during the war. The commission found that "approximately one-half of the employees were subjected to the twelve-hour day" and that "approximately one-half of these in turn were subjected to the seven-day week." If anything, the report concluded, hours worked per week in the steel industry were longer in 1920 than before the reforms of the prewar era.[7]

Meanwhile, Paul U. Kellogg, editor of *Survey* and one of Dickson's allies in the earlier struggle for hours reform, went into action. In his capacity as a trustee of the Charles M. Cabot Fund of Boston (a foundation established as a memorial to the man who forced U.S. Steel to investigate itself in 1912), Kellogg secured funding for two studies that were influential in finally ending the twelve-hour shift. The first grew out of a meeting of the Engineers' Club of Philadelphia not long after the steel strike. There, following a discussion of techniques by which industries changed from the two-shift to the three-shift system, a group of engineers headed by Morris L. Cooke formed a Committee on Work-Periods in Continuous Industry. The committee asked Cooke to determine what progress had been made by the steel industry in moving from the two-shift system. Needing money to finance the project, Cooke turned to the Cabot Fund and received $5,000, which he used to hire two distinguished engineers, Horace B. Drury and Bradley Stoughton, to do the necessary research.[8]

Kellogg also secured a Cabot Fund grant for another study by John A. Fitch to determine where matters stood with regard to the long shift at U.S. Steel. Eight years had passed since the annual meeting at which the stockholders' committee, appointed at the insistence of Charles M. Cabot, had condemned the twelve-hour shift and seven-day workweek and called for abandonment of both as soon as practicable. This second study got under way on July 15, just ahead of the Interchurch World Movement's report.[9]

Drury revealed his findings first. Addressing a joint session of the Taylor Society and sections of the American Society of Mechanical Engineers on December 3, 1920, he showed, among other things, that the twelve-hour shift was neither inevitable nor necessary. A number of independent American steel manufacturers had successfully adopted the eight-hour shift, he pointed out, and it generally prevailed in the steel mills of Britain, Belgium, France, Germany, Sweden, Italy, and Japan. Possibly only in China and India were hours longer than in the steel works of America.[10] Drury's address was widely quoted and commented on in the months that followed.

Then came Fitch's findings. Kellogg devoted the entire March 5, 1921, issue of *Survey* to the twelve-hour problem. In the lead article, Fitch announced that "the proportion of twelve-hour workers is as great today as it was eight years ago when the stockholders' committee of the greatest of all steel companies brought in its report." [11] In the same issue, Whiting Williams explained how the steel industry in Britain had moved to the three-shift system, and S. Adele Shaw wrote a summary of Drury's findings. Even *Iron Age*, traditional spokesman of the industry, carried attacks on the long day. Its March 10 issue, for example, included an article by Williams in which he quoted the reaction of an English steelworker to the long workday in America: "No! Wot! Twelve hours and no time out for breakfast or dinner! —In America? And seven days a week! Well, Rule Britannia! I supposed we was bloody well the lawst! Blime, yer don't sye!" [12]

The mounting pressure led Gary once more to adopt evasive tactics. On March 8 he announced that U.S. Steel had ended both the seven-day workweek and the twenty-four hour turn in all its plants and that a committee of subsidiary presidents was studying the twelve-hour day question. Because the group had not yet reached a conclusion "entirely satisfactory to all of them with respect to some of the features involved," the committee would not report for a month or so. (Dickson wryly observed that this was the third time that Gary had proclaimed an end to the seven-day workweek at the corporation.) One month later the chairman told stockholders at the annual meeting that the twelve-hour shift and the seven-day workweek might never be entirely eliminated from the industry. In May he reported that the committee of subsidiary presidents had "expectation of making the elimination of the twelve hour day during the coming year." Conditions in the industry were depressed at the time and a number of independents had already adopted the three-shift system in order to spread work. By 1922, however, prosperity had returned. A growing labor shortage left the previous year's "expectation" unrealized at U.S. Steel.[13]

Unfortunately for Gary, labor problems in general had taken on political significance by early 1922. Unrest among coal and railway workers and continued bitterness among steel workers threatened to produce the most serious labor crisis in the United States since the 1890s. With contracts in coal about to expire and with growing dissension on the railroads, the Warren Harding administration repeatedly tried to bring employers and employees together around the bargaining table. No one worked harder for harmony than Secretary of Commerce Herbert Hoover. Although he believed that the government should assist management and labor in achieving peace, even by occasional proddings if necessary, voluntary solutions were preferable to those imposed by government.[14]

Apparently it was Kellogg who initiated the steps that brought the administration directly into the hours conflict in the steel industry. Kellogg's fellow trustees of the Cabot Fund had only reluctantly supported his moves to reopen the long-day question and were content to stop once Drury and Fitch had made their reports and Judge Gary had professed the good intentions of the industry. Having been fooled by Gary's promises before, Kellogg wanted to apply more public pressure. For this purpose he once again enlisted the aid of Samuel McCune Lindsay, the Columbia University economist. Lindsay, who commanded the respect of the Cabot Fund trustees, was commissioned by them to organize pressures on Gary from within the business community. A "host in himself," Lindsay seemed to know everyone worth knowing. In addition to a wide acquaintance among academicians, engineers, and clergymen, he also had contacts with important New York business and banking figures. More important from Kellogg's point of view, Lindsay had been an influential adviser to leading Republicans during the 1920 campaign and in 1921 had served on Secretary Hoover's conference on unemployment. In the hours battle, Lindsay served as chief coordinator, collecting and distributing information, arranging meetings, and using his contacts in government and business to put pressure on Gary and the steelmen as needed.[15]

On March 14, 1922, Lindsay called on Dickson to discuss a White

House conference on the twelve-hour day. The economist hoped to interest Harding and Hoover in the project and believed that Judge Gary was favorably inclined. Apparently Lindsay thought the conference would provide a means by which Gary could save face as he retreated from his earlier position. "I hope so," Dickson noted, "but 'hae me doubts.' "[16]

Lindsay, meanwhile, enlisted Hoover's support. As the secretary tried to persuade Gary to act voluntarily before Congress involved itself in the matter, Lindsay pointed out to Gary the moral, economic, and public relations benefits that would flow from the reform. Hoover anticipated that Gary would make a favorable statement on the question at U.S. Steel's annual meeting on April 17. To assure that result, he proposed that President Harding send the Judge a letter designed "to stir his imagination." Hoover enclosed a draft for Harding to sign and apparently planned to make the letter public. Although Harding liked the idea, he declined to make his letter public and changed the Hoover draft to say simply that he personally would be very pleased if the twelve-hour workday in steel were abolished.

The steelmaster disappointed both Harding and Hoover. U.S. Steel, Gary told the corporation's stockholders, had ended the seven-day workweek and the long turn during the past year and since October 1920 had reduced the percentage of twelve-hour men from 32 to 14 percent. Those still working the long shift, Gary explained, were in continuous operations where machinery had to be kept going without interruption. "There is no other practicable way," he insisted. He reiterated all the old excuses: corporation officials would like to eliminate the long shift, but the workmen wanted to labor twelve hours a day in order to earn more, and, actually, the long shift was no hardship because twelve-hour men worked only about half the time during the shift.

To "outsiders" who urged the company to force the men to accept shorter hours, Gary declared, "We do not believe in that doctrine. We believe the workmen are entitled to be consulted. . . . When it comes to the welfare of the workmen themselves, we think they are entitled to receive special considerations." U.S. Steel and its subsidiaries were "doing business at the old stand," he assured his listeners. "Our competitors believe in the twelve hour day, and although they have been more than once requested to express an opinion, so far as we know, they have not yielded to the desire of the lecturers and some publishers, well-intentioned persons, to reduce the hours contrary to the wishes of the men."[17]

To keep the issue alive, Lindsay pointed out to Hoover that the

American Iron & Steel Institute would be meeting at the end of May. He proposed that Hoover urge President Harding to call a White House conference of steelmen and of bankers and businessmen with influence in that industry. The idea was adopted, and on the evening of May 18, forty-one steel executives and business leaders dined with the president.[18] At Lindsay's suggestion, Dickson was invited but could not attend because he was in St. Luke's Hospital in New York "awaiting a possible major operation."[19] As he lay on his bed, "hoping against hope that I might be able to attend," Dickson wrote out what he would say if asked to speak. "Mr. President: Doubtless you have heard the question, 'How should you cook a rabbit?' and the answer, 'first catch your rabbit.' This advice has been accepted generally as being the first step in the process; but it is not. The first step is to *want* rabbit. When the gentlemen seated around this table want to abolish the twelve hour day, there will be no real difficulty in doing so."[20] It was one of the great disappointments of Dickson's life, he later wrote, not to be in at the death of the inhuman system he had fought so long. His forced absence was not his only regret. "What a pity Judge Gary and his associates have failed to do voluntarily that which public opinion will eventually force them to do," he observed.[21]

The after-dinner discussion at the White House on May 18 brought no sudden capitulation of the industry. Harding announced that his purpose in calling the session was to remove the "twelve-hour day blight" from the industry. Armed with charts, graphs, and statistics, Hoover then set forth the case against the long shift, arguing that any increase in costs brought about by the change would be offset by gains in worker productivity. Gary and Schwab replied that Hoover's views were "unsocial and uneconomic." After considerable acrimony and heated debate, Harding intervened to propose that Gary appoint a committee to study the matter further and make recommendations. The steelmen agreed, the party broke up, and Gary named himself to head the study committee. Hoover, "much disheartened [and] in less than a good humor," went directly from the table to news reporters assembled outside the White House and revealed the purpose of the meeting.

Hoover, Lindsay, Kellogg, and other reformers "kept the pot boiling in the press." In his role as president of the Federated Association of Engineering Societies, Hoover arranged for a committee of engineers, headed by Horace Drury, to bring together in a single study all recent engineering reports on the eight-hour day, especially as related to steel production. The committee found that the twelve-hour shift was both economically unsound and socially unwise, and that the eight-hour system was advantageous to both employees and employers. The changeover to the three-shift system in steel, the engineers estimated, could be accomplished without raising costs more than 3 percent. Hoover drafted a forward to the study, which Harding signed, expressing satisfaction that the findings of the engineers coincided exactly with his (the president's) own sentiments. "[T]he twelve hour day and the type of worker it produces have outlived their usefulness," Harding declared. "The old order . . . must give way." [22]

Meanwhile, Lindsay busied himself trying to arouse inside business pressures on Gary and the steel manufacturers. Now a trustee of the Cabot Fund, Lindsay helped find money for publication of the engineers' study. Then, early in 1923, he persuaded Dwight Morrow to give a luncheon for Owen D. Young, Robert Lamont, Eugene Grace, and a number of other prominent business and financial leaders. Hoover was guest of honor, and in his address summarized the findings of the engineers once more and warned that if the steel industry did not act quickly, "they would face Congressional legislation, which to his mind and theirs, was undesirable." [23]

The committee headed by Gary delayed its report for a year. Among other things, Gary contacted German steelmasters to discover the impact that the eight-hour shift had had there. Although their efficiency studies had led them to believe the three-shift system a mistake, they gave Gary both positive and negative evaluations. The positive findings were deliberately included "to side track" the Americans "and raise their costs." The German steel industry returned to the twelve-hour shift at the end of 1923.[24]

At a meeting of the American Iron & Steel Institute on May 25, 1923, Judge Gary, having just returned from a Mediterranean cruise, told the steel executives that there was "only one way of fairly and finally settling any controversy," and that was "in consonance with the principles of the Christian religion." Before completing his remarks, an indisposition forced him to leave the platform and the task of reading the committee's report fell to Schwab. Once more, reformers were disappointed. The committee rejected any move to the three-shift system at the present time. "Had the incident [Gary's sudden illness] happened in mediaeval times," the *Baltimore Sun* commented, "there would have been many to see in it the finger of God, halting a hypocrisy, perhaps unconscious, which might have been characterized as near to blasphemy." (Dickson underscored the word "hypocrisy" on his copy of the clipping.) The Gary committee argued that the workers preferred the long shift. "It is asserted with confidence that there is less fatigue from the work of the twelve hour day in the steel industry, than pertains to the large majority of the eight-hour men." The committee found that the changeover, even if possible, would add about 15 percent to the cost of steel production and would require no fewer than sixty thousand additional workmen. Because Congress had seen fit to limit immigration, the committee complained, the industry currently was having difficulty manning two shifts.[25]

Many people apparently thought that the year's delay had been a face-saving device for the industry. The flat refusal, though labeled by Gary as applying only to the present situation, outraged the press, the clergy, and business and labor leaders alike. "Gentlemen of the steel industry," declared B. C. Forbes in the *New York American*, "you will have to try again. Nothing is ever settled until it is settled right." [26] Horace Drury, writing to Morris Cooke, denounced Gary's *"old stuff."* Did it cost Henry Ford, whose River Rouge steel mills operated on the three-shift system, 15 percent more to turn out pig iron than it cost U.S. Steel? As for the labor shortage, the industry by stalling the reform for two years had missed the opportunity to take advantage of an oversupply. "No labor shortage is so severe," Drury observed, "but what a gradual movement from two to three shifts could proceed." [27]

Secretary Hoover drafted a response to the committee's report, which Harding signed almost without change and sent to Gary on June 18. Expressing great disappointment, the president hinted at the possibility of governmental intervention when he said that he still hoped that the matter could be solved within the industry itself. Accepting the argument that labor was in short supply, Harding wondered if it would not be possible for the industry to pledge to make the changeover just as soon as workmen became available. "Such an undertaking," he concluded, "would give great satisfaction to the American people" and would "establish pride and confidence in the ability of our industries to solve matters . . . so conclusively advocated by the public." [28]

Although Gary favored holding firm, he called a meeting of the directors of the Iron & Steel Institute. The steel executives faced a difficult choice: they could admit, as Harding urged, that the twelve-hour day was an evil and promise to end it as quickly as possible, or they could risk further hostile public reaction and possible legislation by refusing to yield. The directors chose the former, pledging to secure "a total abolition of the twelve-hour day at the earliest time practicable." [29] If Gary thought that he was once again buying time, he was mistaken. Harding, then on his ill-fated western tour, announced in a public address at Tacoma, Washington, on July 5, that the steel industry had pledged to end the long shift. To insure that there would be no subsequent backsliding, Secretary Hoover made the president's correspondence with the American Iron & Steel Institute public.[30] In the weeks that followed the press kept the issue regularly before the public.

On August 3 the front pages of the nation's newspapers carried the story of Harding's death the evening before. By coincidence, on their inner pages, the same papers reported Judge Gary's announcement that U.S. Steel was beginning the elimination of the long shift in all of its plants. Gary's surrender was not graceful. "Our men will be bitterly disappointed," he told newsmen. "The workers in the steel industry of America are an ambitious group. Nine out of ten of them would rather work seven days a week. And they make more money in the twelve-hour day than they will make in the eight-hour day, in spite of the raise in salary. We can only hope that they will take it good-naturedly. . . . This sudden change will come as a blow to them." [31] The chairman's lack of touch with reality moved Dickson to verse:

The Steelworker's Lament

Mike Miller, on a summer day,
Raked the furnace, hot as say—
The hinges of old Satan's lair;
And, as he raked, he tore his hair.

Alas for me, misfortune lowers;
I'm not allowed to work twelve hours.
No more at five, may I arise,
And rush with half awakened eyes

To toil once more, midst heated bricks
And molten steel, from six to six.
O woe is me, just eight brief hours,
And then, to sunshine, air, and flowers,

And wife and children I'm condemned.
I'm in despair. My life is hemmed
About with such restrictions, I
Might as well give up and die.

O Judge, don't leave us in the lurch.
Next thing they'll drag us off to church,
And make us list to hymns and psalms,
On Sunday morn, instead of damns,

And other expletives with which
Our bosses their commands enrich.

Our lives were quite ideal, we
Were very happy. Let us be.

And tell those poor romantic fools,
That if these new eight hour rules
Are forced on us, we'll up and strike.
We can't abide our homes, we'd like

To live in a stockade, and then,
The Judge could, every night at ten,
Tuck those in bed who had a right
To this brief rest; kiss them goodnight,

And ask them if they'd been good boys.
But no such luck; Utopian joys
Are not for us. Misfortune lowers.
We're not allowed to work twelve hours.[32]

Gary's hope that the workmen would understand and not revolt against the shorter workday was realized. His predictions that the changeover would require sixty thousand men and add 15 percent to costs were both proved wrong. The industry experienced little difficulty in finding men, and costs rose about 5 percent. *Survey*, in January 1927, examined the "havoc" that Gary had predicted in 1923. U.S. Steel had paid extra dividends of three-fourths of 1 percent on its common stock in 1923 and 2 percent in 1924. In 1925 the corporation raised its regular dividend from 5 to 7 percent and in December 1926 announced a 40 percent stock dividend.[33]

In October 1923, six months after abandoning the long shift, Gary once again defended the twelve-hour principle at a meeting of the American Iron & Steel Institute. Dickson, who was present, afterward recorded in his diary, "The Steel Corp. has finally adopted my views on the seven day week and the 12 hour day but only after being discredited before that great court of final resort, 'Public Opinion.' And I—but what's the use?" [34]

* * *

Vindication of his belief that steelworkers must be given a greater voice in determining the conditions under which they labored came more slowly and in a form not to Dickson's liking. Once Midvale adopted his ERP in 1918, Dickson urged it on Gary, who only laughed the idea to scorn. After the 1919 strike, Dickson again pressed for employee representation throughout the industry. "I believe most earnestly that the basic idea . . . offers the only solution of the proper relation of capital and labor, and wish that Judge Gary could be influenced to adopt the plan," he wrote to Charles L. Taylor. "I cannot believe that the present status of labor can be maintained permanently by the United States Steel Corporation, however benevolent in intentions and practice. In effect a few men, (really three men, i.e. Judge Gary, Mr. [George F.] Baker and J. P. Morgan Jr.) have absolute control over half of the steel industry and all persons affected by it. This is repugnant to the spirit of our institutions." [35] Dickson, of course, lost whatever influence he may have had in the industry once Midvale was absorbed into Bethlehem Steel in 1923. Even his creation, the Midvale Plan, was replaced by Bethlehem's own company union.

Collective bargaining, whether by company union, employee representation, or regular union, made little headway in the steel industry prior to the New Deal. Once collective bargaining became mandatory in 1933, Dickson watched, satisfaction mixed with disgust, as the industry tried to set up company unions so as to forestall regular unions and at the same time planted spies among the workers and purchased arms to battle them. Dickson could not resist writing to the chief officers and directors of U.S. Steel on their folly. "The organization of company unions was a step in the right direction but unfortunately it was delayed too long, with the inevitable result that the good faith of the Corporation was questioned by the general public, the Federal Government, and by many of the workmen." As for collecting munitions, he wrote to a vice-president about having watched the battle of Homestead forty-five years before. "In the name of God," he pleaded, "have the men in control of the Corporation no memories? 'Whom the gods would destroy, they first make mad.' " [36]

Writing to President Franklin D. Roosevelt in June 1934, Dickson advocated not only employee representation in industry, but also labor representation on corporate boards of directors and profit sharing by employees. "Such a system should be established," he wrote, "not as an act of grace on the part of the employer, but as a concrete recognition of the identity of interest of these three partners [capital, management and labor]." At the same time, Dickson condemned legislation that tended "to perpetuate the vicious idea that capital and labor are necessarily arrayed in hostile camps; and that their normal relations are those of warfare, or, at best, an armed truce." [37]

The sudden capitulation of U.S. Steel to unionization in March 1937 surprised both labor and capital. Dickson was despondent. "Recent events indicate that my hope of an industrial democracy must be deferred indefinitely; and that we shall have instead, American industry entrenched in two hostile camps—an autocracy of capital facing an autocracy of labor; operating under a temporary truce—*but resting* on their arms! What a pity, that the monumental stupidity of men in high places, who have mistaken mere inertia for true conservatism should have brought such a condition to pass." [38]

* * *

Dickson's last years were frequently burdensome to him. His relative poverty and inability to find a satisfactory outlet for his drive and energy created intense unrest. He dabbled in real estate, wrote pamphlets and a newspaper column, he sent letters to congressmen and even to the president, and he pestered officials of U.S. Steel with long letters explaining the accuracy of his past predictions and proffering advice. For a while he advertised himself as a speaker before chambers of commerce and similar civic groups, but with little success. During his last years he spent much time in going over his files of letters and news clippings for the 1910–1923 period, making copies, inserting comments, and drafting memos. He sketched out an autobiography, but bogged down in his writing when he reached the chapters dealing with his battles at U.S. Steel. In 1938 he compiled a *History of the Carnegie Veterans Association*.

During the early 1930s, Dickson devoted himself to plans for ending the depression. Among his earliest formulations was a proposed third-party platform for the election of 1932. Blended together were a curious mixture of radical and, for Dickson, uncharacteristically reactionary planks: repeal of prohibition (though he himself was a lifelong teetotaler); repeal of the antitrust laws (to be followed by the enforced consolidation of industries affecting the public interest and the encouragement of voluntary consolidation of other major industries); complete planning of the economy by a small body representing producers, consumers, and laborers; protection of the railroads from their competitors; limiting the franchise by educational and property qualifications; restoration of the indirect election of senators; taxation "a la Henry George" of all reserves of natural resources; tariff reduction; a six-year term for the president coupled with ineligibility to serve a second term until the lapse of at least six years; recognition of the Soviet Union; joining the League of Nations and World Court; and cancellation of war debts.[39] By February 1933, Dickson had dropped most of these planks and was concentrating on the consolidation of industries and the detailed planning of the economy by producers, consumers, and workers.[40]

After 1934, Dickson returned to a scheme for profit sharing that he had originally formulated in 1931 and that he thought would improve the nation's economic climate. According to his plan, capital, management, and labor were to become truly equal partners in the industrial process. Stockholders were to receive as "wages for capital" a dividend equal to the prevailing interest rate, plus an additional dividend (equal in amount to the first) as insurance against depreciation of investment. Managers and workers were to be paid salaries and wages in accord with amounts being paid generally for comparable work and were to be protected by old age and unemployment insurance and by workers' accident compensation to which they would contribute. These "wages" for the three partners were all to be treated as costs of production. Any profits beyond expenses were to be divided proportionately among the stockholders, managers, and workmen according to the contributions of each to the partnership.

Outlining his scheme in pamphlet form, Dickson sent it broadcast to businessmen and politicians.[41] It drew little enthusiasm. "Your idea of profit sharing among the three partners to industry," J. P. Morgan replied, "is a very interesting one, but . . . in the past five years it would have been impossible under any circumstances to share profits, because there were none." E. J. Buffington pointed out that "certainly employees do not want to bear any part of industry's losses incident to adverse business conditions, . . . and without such an arrangement, there, of course, can be no real partnership." [42]

Writing in his diary in April 1937, Dickson noted that although he had passed "the scriptural limit of 'three score years and ten,' " and his days were " 'in the sere and yellow leaf,' " his interest in the labor question persisted. In fact, that life-long passion had nearly spent itself. After his retirement contacts with men of affairs became less frequent. Moreover, one by one the men with whom he had argued the issue passed from the scene: Judge Gary in 1927, Dinkey in 1931, Corey in 1934, and Schwab in 1939. Dickson himself lived on till January 27, 1942.

* * *

Throughout his lifetime, Dickson somehow had never been wholly at ease with the steel executives among whom he lived and worked. If, in the beginning, the driving ambition they all shared was tempered by nobler and selfless aspirations, most of his associates had soon abandoned idealism for practical ends. Dickson remained "romantic" (as he termed it) to the end. He was unable in his thinking to divorce success in industry from concern for the workingmen who made success possible, and he stubbornly held that in the quest for profits there must be restraint in the form of fair play for competitors and justice for employees. Dickson argued that power could be shared, at least in part, with workmen whose lives were so deeply involved in industrial decisions.

Shaken by the triumph of profits over justice in the battle for hours reform at U.S. Steel before the war, Dickson clung to his beliefs. At Midvale he renewed the struggle, this time seeking to give workers some control over work conditions. When that battle, too, was lost, he refused to believe that he was mistaken. Instead he blamed defeat on the narrowness and lack of vision of the other steelmasters. Their single-minded quest for power and profits, their lack of interest in intellectual or cultural matters, and their unwillingness to reflect on the consequences of their behavior offended him.

Formal dinners of businessmen, he complained, were "long on food and drink and short on brains." "I am constantly surprised at the low level of mentality displayed at these public dinners," he observed on another occasion, "silly compliments, bromidic utterances, and so seldom anything really gripping." Following a luncheon with the chief officers and directors of Midvale Steel in March 1923, Dickson commented that "at such a gathering one would expect either serious worthwhile discussion of topics of the day or real wit. We had neither." [43]

The entertainments put on by his peers were even worse, Dickson believed. An affair given by William E. Corey was typical. "There were present fifty nine people and about thirty entertainers," Dickson noted, "the latter mostly dancing girls. The list of guests included some of the most prominent business men and bankers in N.Y. WEC told me that over 200 bottles were consumed. I left about midnight. While there was nothing unusual in the entertainment, I marvelled that men of such high attainment could find any real satisfaction in pleasures on such a low level—merely consuming food [and] liquor and looking on while a lot of brainless girls danced before them. There was nothing indecent in the entertainment but the whole thing to me was utterly banal." [44]

Dickson's private evaluations of the great leaders of the industry (usually written in his diary at the time of their deaths) were harsh. "Death of H. C. Frick announced. A man I never admired. He was a cold task-master and was primarily responsible for the continuance of the twelve hour day and the seven day week in the Carnegie Steel Co. and U.S. Steel Corporation Mills. He will no doubt leave a large fortune. In my opinion the manner in which a man makes his money is more important to his fellows than the manner in which he disposes of it. He debased his fellow men in the acquiring of his money." [45]

Carnegie, whom Dickson admired, had received an only slightly more favorable rating on his death a few months earlier. "To future generations must be left the task of estimating this notable man. . . . In the use of his great fortune he has strenuously endeavored to help his fellow man. The methods by which it was acquired were not so commendable, especially his attitude toward labor as shown by the twelve hour day and seven day week in his steel mills." [46]

Although Judge Gary often exasperated him, Dickson was fair in appraising him. Of a speech by Gary before the American Iron & Steel Institute, Dickson observed that it was "the usual balanced series of bromidic utterances reminding me of Oliver Wendell Holmes' 'Katydid.' 'Saying undisputed things, In such a solemn way.' " Yet when Gary died in 1927, Dickson referred to him as "a wise administrator, handling the peculiar industrial and economic problems of our age with far-seeing statesmanship. In this respect a wiser man than Mr. Carnegie. Unfortunately, for me (and I think also for him)," Dickson continued, "in my ten years of intimate association with him . . . my influence was not strong enough to prevail over that of my reactionary associates such as Schwab, Corey and Dinkey." [47]

Occasionally Dickson drew satisfaction from the confession of an associate that he had been right all along. At Dinkey's funeral in 1931, Schwab made such an admission. "Billy," he said, "when I review our past history, I am very much ashamed of the way we treated our labor. In my opinion, the working men have higher standards than the average of our class." [48]

Temperament and zeal had set Dickson apart and made him a gadfly of the steel industry for a decade and a half. Despite his efforts, however, he could take little credit for ending the twelve-hour shift or the seven-day workweek. Others won those battles. His efforts to introduce "industrial democracy" failed because neither he nor the representatives chosen by the workers had any idea as to how to convert the hastily imposed system into a genuine partnership of capital and labor; the cynicism of Dickson's superiors would have made the attempt futile in any event. In the end, Dickson's career measured the limits of what a single, dedicated reformer, crusading for a just cause, could accomplish. Even more forcefully, his experiences revealed how long a single, determined opponent of reform such as Gary, armed with power, could thwart change, however necessary, however reasonable, and however strongly the change was supported by public opinion.

* * *

What significance, if any, did the events here chronicled have in the larger context of the restructuring of American industry between the 1890s and the 1920s? One matter of importance is perspective. To those who look back on the period, it is clear that the main thrust of the changes in steel, for example, was in the direction of the complete integration and rationalization of manufacturing and marketing. That reordering, in turn, aimed at a greater stability of supply, of prices, and of profits. This study, however, has tried to view events front-end-to, that is, as they unfolded rather than in retrospect. As a result, the primary thrust and the objectives of the reshaping of the industry seem somewhat less obvious. What happened in no sense had to happen. There were alternatives to the courses pursued, and they suggest that other lines of development were possible. Similarly, by emphasizing the roles of individuals in bringing about the restructuring, rather than treating steelmasters as a group, the interplay of ideas, personalities, and ambitions in the shaping of events becomes more evident.

The men who wrought the changes in steel, for instance, obviously did not share a common vision or agree on what it was they were trying to accomplish. A mingling of individual ambition and corporate welfare (with an occasional nod to humanitarian considerations) shaped their actions. In effect each sensed only part of the new order that was emerging, even as each also clung to a part of the old order that was dying. Their diverse objectives, personal and collective, in part caused, and in part were caused by, a power struggle to gain mastery over the principal firm in the industry. That contest resolved, the victor proceeded to impose his views on U.S. Steel and to a lesser degree on the industry as a whole. The so-called managerial revolution had brought new types of leaders to the top, but, for the time being at least, control of big business was no less authoritarian than it had been under the industrialist-owners of the previous era.

But even as the new leaders established their hold, a fresh challenge appeared: how to reconcile their tight, essentially autocratic control with America's long-standing liberal-democratic tradition?[49] This question embraced both the "corporate problem" and the "labor problem," matters that would be of major concern to the nation for the next half-century. The problem of corporations included the relationship of managers to one another and to the stockholder-owners who technically employed them. It also involved the degree to which corporations were accountable to the general public for their conduct. The labor problem, until the appearance of big unions after 1933, centered in the inconsistency between the large earnings of industry and the deplorable conditions under which workers lived and labored. Also raised was the question whether or not workers were to have a voice in fixing the terms of their own employment.

The lack of a common vision among those who transformed the steel industry after 1890 was reflected in their views on integration. Carnegie, who years before began the large-scale integration of steel production, had started with no overarching purpose of rationalizing the business. Rather, his quest for ways to lower production costs so as to best his competitors led him into integration. Maximum cost efficiency in the manufacture of basic steel, he discovered, required that he control both the production of his own raw materials and the transportation system needed to move them to his mills. Again, near the end of his career as a steelmaster, Carnegie considered forward integration into the fabrication of steel. His reasoning was the same: not to create a reasoned and orderly process so much as to counter the actions of competitors.[50]

Schwab, like Carnegie, was production-oriented. In his celebrated after-dinner address that helped to bring U.S. Steel into being, he talked of completing the integration of steelmaking from raw material to finished product. Further reductions in the cost of production through improved technology and the harder driving of workers in the mills were unlikely, he believed. Any new savings would have to come from the further consolidation and rationalization of the industry. That would permit individual plants to specialize in single products. Duplication of function, such as each firm maintaining a separate sales force, could be eliminated. New mills could be located in or near the markets they would serve, thereby greatly reducing the crosshauling of steel; customers would simply be served from the closest plant. Schwab saw ever more efficient production and the development of an orderly distribution system in an ever expanding market as the justification for further integration and consolidation.[51]

By contrast, Gary (and Morgan who backed him) anticipated an increasingly stable market in which the rapid expansion of the past would no longer prevail. The Judge favored consolidation because it would make possible a more predictable marketing of steel. Properly controlled, the constant drive for technological innovation, reduced costs, slashed prices, and intense competition would give way to harmonious cooperation among steel firms. Production, prices, and profits would all be stabilized once the quest of individual companies for short-range advantage was replaced by an industry-wide goal of long-range certainty. Planned production, with output geared to market demand, offered more reliable earnings than did the unruly competition of the Carnegie era.

Steelmakers generally were slow to grasp the full significance of the dominant market position of the newly formed steel trust or the advantages of noncompetitive pricing. Only after being driven from command at U.S. Steel and being instructed along with the other independent steelmasters in the dangers of entering into competition with U.S. Steel did Schwab, for instance, become a champion of "harmony" within the industry.

Dickson, similarly trained in the competitive school, never accepted Gary's commercial policy. It was an "unworkable theory," he believed, because there was no lawful way to fix prices effectively. Moreover, when U.S. Steel ceased to behave in a competitive manner, it lost most of the advantages that it had realized through integration and size. As an official of the corporation, Dickson opposed Gary's policy, and as an independent steel producer after World War I he led an unsuccessful revolt against rigid prices. At the same time, Dickson rejected the ruthless competitiveness of the Carnegie era. Competition must continue, he insisted, but must be fair, restrained, and nonpredatory. Exactly what that meant and how it differed from the "wrongheaded" policies of Carnegie on the one hand and Gary on the other, Dickson never made clear. Gary won out and noncompetitive pricing held sway. It is unlikely, however, that the subsequent history of either U.S. Steel or the industry would have been the same if competition, even in the restrained form recommended by Dickson, had continued.

As for the administration of modern industry, the steelmasters were in closer agreement. Although ownership had passed from owner-operators such as Carnegie to armies of stockholders, the managers employed by the corporations and holding companies exercised as absolute control over their firms as Carnegie and his contemporaries. It was true that, until the lines of authority were clearly established, U.S. Steel was run collectively by its president, executive committee, and the chairman of its board of directors. But this was a temporary expedient, not a situation favored by any of the leaders. Neither President Schwab nor Chairman Gary regarded collective (much less democratic) leadership as desirable. When Gary won, Corey and Dickson fought a rear-guard action in defense of decentralized control. This was more to preserve their own dwindling powers and to curb Gary, however, than from a conviction that less centralized control in itself had merit.

The roots of the elaborate techno-structures that govern modern American industry can be traced to the era under study.[52] In steel, however, those developments lay at least a generation in the future. Gary, to be sure, made much use of the large bureaucracies maintained by U.S. Steel and its various subsidiary companies. All important decisions, however, ultimately were made by him. Overmuch has been made by some scholars of the power of Carnegie-trained subsidiary presidents to thwart Gary's policies, particularly in the early years of

U.S. Steel. For example, they argue that Gary favored the abolition of all unnecessary Sunday labor within the firm and its subsidiaries but could not win over the presidents. These executives did oppose the reform and failed to obey the rule against such labor, which they themselves had adopted in 1907.[53] From Dickson's experience in testing the rule, however, it is evident that Gary did not regard the reform as particularly important and in the end sided with the steelmakers against it. More to the point, when Dickson, backed by outside reformers, subsequently pressed the issue, Gary ended Sunday labor by fiat without so much as consulting the allegedly powerful subsidiary presidents. That order was obeyed, moreover, until labor demands occasioned by World War I made it possible for Gary and his subordinates once more to let the matter slip.

Gary's predominant position in the industry as a whole dated from his "dinners" and his enforcement of noncompetitive prices during the panic of 1907. Formally organized into the American Iron & Steel Institute after 1910, the industry leaders thereafter accepted Gary as their chief spokesman for the balance of his life. He presided over the institute, addressed the group at every session that he attended, named all of its committees, and chaired those that he regarded as important. Whenever the industry needed someone to address the public on its behalf, or to negotiate with the government, Gary stepped forward. Although he did consult with the chief executives of other firms from time to time, Gary's views emerged as the views of the industry and the decisions he made were generally accepted as binding on all. In the matter of the governance of U.S. Steel, Schwab, or even Dickson, as chief executive officer of the firm, would probably not have produced a more democratic rule. At the same time, under Schwab or Dickson is it unlikely that the Gary dinners or the American Iron & Steel Institute would have come into being. It is also possible that the head of U.S. Steel, though still the most powerful of the steelmasters, would have had much less influence over the industry at large.

As Gary's authority at U.S. Steel became absolute, he recognized the need to meet a growing clamor against secretive and authoritarian control of corporations by their hired managers. His solution to this problem, as to many others, was to turn to public relations devices. To create the impression that U.S. Steel was openly and democratically run, he encouraged stockholders to attend annual meetings of the corporation and newsmen to cover the sessions. He met with reporters to answer questions and to issue news releases. Frequently he disarmed critics of the corporation and industry by appointing committees to investigate and to report on complaints.

Long before 1901, however, stockholders of large corporations (not to mention firms controlled by holding companies) had lost any meaningful control over boards of directors and corporate managers. At U.S. Steel's annual meetings, for example, Gary always exercised proxies for more than 95 percent of the voting stock. That a minority stockholder in 1911 forced the corporation to investigate its own labor policies was not an indication of stockholder power; it was an example of Gary's deliberate cultivation of the illusion that stockholders had an important voice in company affairs.

Similarly, Gary paid lip service, but little else, to the idea that U.S. Steel and the American steel industry were subject to governmental regulation in the public interest. However, under Gary's leadership the industry engaged in pooling to maintain stable prices. Because he was careful not to call the arrangement pooling, Judge Gary insisted that the practice, therefore, was not illegal. In a statement to the Stanley Committee in 1911 (which revisionist historians have taken at face value and frequently cited as evidence of the Judge's true goals), Gary declared that he believed the steel industry "must come to enforced publicity and governmental control."

Q. You mean governmental control of prices?

A. I do; even as to prices. . . . I would be very glad if we knew exactly where we stand, if we could be freed from the danger, trouble, and criticism by the public, and if we had some place where we could go, to a responsible governmental authority, and say to them, 'Here are our facts and figures, here is our property, here our cost of production; now you tell us what we have the right to do and what prices we have the right to charge.'

Q. Is it your position that cooperation is bound to take the place of competition?

A. It is my opinion.

Q. And that cooperation therefore requires strict governmental supervision?

A. That is a good statement of the case. I believe it thoroughly.

Q. And that the Sherman Act is archaic?

A. It is.[54]

Gary's sincerity in all this must be weighed against his conduct whenever government, for whatever reason, attempted to invade the domain of steel. As has been repeatedly shown, he resisted the slightest restraints on corporate independence. Even during so great a national emergency as World War I, he fought against governmental controls over steel prices and interference with wages and hours. These acts, far from supporting the contention that Gary welcomed governmental enforcement of stable prices, suggests instead that what he said must always be measured against what he did and that he would go to almost any length to evade governmental regulation.

The nature of labor reform in the steel industry in the years under consideration was determined more by Gary's power at U.S. Steel and his influence over the industry at large than by any other single factor. At the same time, two prevailing conditions influenced both the chairman's thinking and that of the other steelmasters. The first was labor's weakness vis-à-vis the companies. Disorganized, lacking in wealth, influence, or power, the steelworkers could not successfully resist encroachments on their union by the companies, much less wring concessions from them. On the other hand, the reformist spirit of the time demanded improvements in the conditions of labor. Middle-class and professional do-gooders, organized, outspoken, and influential, could not safely be ignored.

Schwab, the Corey brothers (William and Fred), and Dinkey, so far as possible, denied the right of outsiders to meddle in employer-employee relations and opposed any concessions to labor. Like their mentor, Carnegie, all four saw labor as a cost of production to be driven to the lowest possible point. Schwab, in the manner of his teacher, disguised his essentially antilabor views behind a cheery countenance and a hearty camaraderie with the workers. The Coreys and Dinkey, in the tradition of Henry Clay Frick, remained cold and aloof and in private spoke of laborers with contempt.

Gary, far more sensitive to public opinion, recognized the need to appease the demands for reform. He always spoke respectfully of workers and their rights and professed a concern for their well-being. The welfare capitalist program that he adopted aimed chiefly at warding off and dampening down criticism, and, as might be expected, impressed reformers more than it did laborers.

Dickson was neither unaware of the value of good public relations nor of the necessity for concessions to public opinion. Unlike Gary, however, he saw the reform climate of the prewar years as an irresistible force of history to be accommodated and not as a passing mood to be weathered. True leadership, he believed, required that businessmen run ahead of the tide, yielding early and voluntarily to those reforms that the public ultimately would force them to accept. Dickson was no simple opportunist, however. His professions of humanitarian compassion were real. In contrast with his colleagues, many of whom had also come up through the mills, Dickson had not allowed his improving station in life to blunt his outrage at work conditions he had experienced. He heartily supported Gary's concessions to labor but saw the need to press beyond mere paternalism—farther, as it turned out, than Gary was willing to go.

Of all the steelmasters, only Dickson warmed to labor unions, and he only temporarily at the close of the war. He was also the only one to accept the notion that workers should have a voice in determining the conditions under which they labored. Schwab, William Corey, and Dinkey all too clearly remembered the days when the Amalgamated had successfully blocked proposals to improve the efficiency of the mills. Once the union was crushed at Homestead, these men opposed any concessions to labor organizations, even those set up and controlled by the companies.

Gary's opposition to any form of collective bargaining had little to do with direct experience with unions. The Amalgamated's strike against U.S. Steel in its first months of existence greatly irritated him, to be sure. More important, however, unions or other forms of collective bargaining would have posed a threat to Gary's absolute authority and would have introduced an element of uncertainty and instability to the industry. The chairman's one concession to worker participation in managerial matters was to suggest that those interested buy stock in the corporation. That was safe enough. Few workers could afford stock, and shareholders, in any event, exercised no effective control.

Unfortunately for Gary, not many people were fooled by this gimmick nor by his various welfare capitalist schemes. Charles Gulick, a contemporary scholar of U.S. Steel, after careful study summarized the corporation's labor policies as "paternalistic and autocratic": its welfare program was paternalistic; its handling of hours, wages, and grievances autocratic. U.S. Steel, he charged, had "always subordinated its interest in reforming hours [and wages] to its interest in output and profits." Such improvements as it did introduce in working conditions "were forced by 'outsiders,' a business depression, or outraged public opinion." Dickson, operating from the vantage point of an insider, agreed completely. He had leveled the same charges against the corporation long before Gulick began his study.[55]

Government insistence that war plants resolve labor problems through collective bargaining of some sort gave ERPs a toe-hold in the steel industry. They made no headway, however, despite the fact that Schwab found the ERP at Bethlehem harmless and continued it until the mid-1930s. Only Dickson among the steelmasters thought company unionism had a future.

In industry at large, Dickson found considerable support. Many businessmen outside steel thought ERPs might be a solution to the labor problem. A company union would be run by employees, not "outsiders," and would be subject to the restraining influence of company representatives at its meetings. Furthermore, company unions stressed the harmony of interests between employers and employees rather than an adversarial relationship. Under those circumstances it seemed probable that they might reduce radicalism, strikes, and similar problems. Workers would benefit too, gaining a voice in matters concerning the workplace and having fellow-workers rather than strangers speaking for them. Finally, company unions would be less expensive for the workers because management paid all the expenses.

As matters turned out, employee representation was headed nowhere. This was not inevitable, nor was it evident at the time.[56] Employers, determined to make their employees full partners in a common enterprise, might have turned company unionism into a workable form of collective bargaining. As at Midvale, however, company unions proved to be only another gimmick, a device for buying time, a means of forestalling genuine unions. The historian Herbert Feis as a young man sat in on a company union meeting. He noted a complete lack of enthusiasm by all the participants. "The men are not reaching forward through the plan," he noted, and "the management has ceased to attempt any great achievements through it." [57] Many company unions, again like Midvale's, received mortal wounds during the depression of 1921 when employers arbitrarily cut wages or went through only "the barest pretense of consultation" with the organizations.[58] At bottom, although industrial leaders had greatly increased and concentrated their own power by expanding their firms and consolidating their industries, they saw no reason to diffuse those powers voluntarily by sharing them with workmen for whom they had little respect.

By contrast, Dickson respected the potential of workers in spite of their present condition and was willing to give them a limited role in shaping the conditions and terms of employment. He failed to convince his colleagues in the industry, however, and seems never to have realized how far short the Midvale ERP fell from effectively representing the interests of its members.

For a steel executive, Dickson's views on labor were advanced as late as 1920. But there his thinking froze. Although company unionism and welfare capitalism quickly became irrelevant as solutions to the labor problem, he stubbornly clung to them for the rest of his life. In his failure to continue growing, he was not alone. Most of the industrial leaders of his generation had exhausted their fund of ideas and were soon to pass from the scene. New crises revealed complex new problems for which they had no solutions. The next stage in the development of American industry would be less controlled by businessmen. After 1933, at least for a few years, elected public officials, bureaucrats, and a new force, the leaders of suddenly powerful labor unions, would share in directing the course of events.

NOTES

LIST OF SOURCES

INDEX

Notes

Introduction

1. Louis Galambos, "The Emerging Organizational Synthesis of Modern American History," *Business History Review,* 44 (Autumn 1970): 279.

2. *Metal Statistics 1978* (New York: Fairchild Publications, 1978), pp. 196–97. After U.S. Steel, the six largest companies and their respective percentages of total raw steel produced in the United States were Bethlehem (13.6%), National (7.2%), Republic (7.2%), Inland (6.4%), Armco (6.4%), and Jones & Laughlin (6%). These seven companies turned out 70 percent of the raw steel produced in the nation in 1977.

3. Martin J. Sklar apparently coined the phrase "corporate liberalism" in "Woodrow Wilson and the Political Economy of Modern United States Liberalism," *Studies on the Left,* 1 (1960): 17–47, esp. p. 41. James Weinstein, *The Corporate Ideal in the Liberal State: 1900–1918* (Boston: Beacon Press, 1968), among others, develops the theme in detail.

4. The best example of the revisionists is Gabriel Kolko, *The Triumph of Conservativism* (New York: Free Press of Glencoe, 1963). Kolko deals with the steel industry especially on pp. 78–81, 114–19, and 170–78. The historical revision of the nature of Progressivism has been complex and in several instances preceded the work of the New Left. The writings of Richard Hofstadter, Samuel P. Hays, and Robert H. Wiebe, to name but three non-New Leftists, contributed significantly to the overthrow of the traditional view of Progressivism.

5. Ellis W. Hawley, "The Discovery and Study of a 'Corporate Liberalism,' " *Business History Review,* 52 (Autumn 1978): 310–11.

6. The best general treatment of the subject is Stuart D. Brandes, *American Welfare Capitalism, 1880–1940* (Chicago: University of Chicago Press, 1976). Morrell Heald, *The Social Responsibilities of Business, Company and Community, 1900–1960* (Cleveland: Case Western Reserve University Press, 1970), pp. 1–53, traces the emergence of welfare capitalist ideas in the business community.

7. Notes for "Steel and Democracy, Memoirs of William Brown Dickson" (hereafter cited as "Memoirs"), Chapter 9, box 2, William Brown Dickson Papers, The Pennsylvania State University Library, University Park, Pennsylvania (hereafter cited as WBD Papers).

8. Ibid. The various chapters of Dickson's "Memoirs" are in different stages of completion. Some are in near-final draft, others in rough draft, some exist as incomplete

notes and others as accumulated documents and raw notes. For some chapters more than one version has been preserved.

Chapter 1. Issues and Antagonists

1. Dickson preserved the telegram, Family Scrapbook 2, box 2, WBD Papers.
2. "Memoirs," Chapter 4, box 2, WBD Papers.
3. James H. Bridge, *The Inside History of the Carnegie Steel Company* (New York: Aldine Book Co., 1903), pp. 281–82, 350, 363–64; Dickson to Charles W. Baker, Nov. 21, 1933, box 5, WBD Papers.
4. The summation of Carnegie's business practices that follows is based on the excellent biography by Joseph Frazier Wall, *Andrew Carnegie* (New York: Oxford University Press, 1970), and Harold C. Livesay's insightful study, *Andrew Carnegie and the Rise of Big Business* (Boston: Little, Brown, 1975).
5. Livesay, *Carnegie and Big Business*, p. 101.
6. For the evolving technology of steelmaking in this period, see David Brody, *Steelworkers in America, the Nonunion Era* (Boston: Harvard University Press, 1960), pp. 1–49. The data on rail production will be found in *Statistics of the American and Foreign Iron Trade for 1912* (Philadelphia: Bureau of Statistics, American Iron & Steel Institute, 1913), p. 106. See also, Brody, *Steelworkers*, pp. 8–9.
7. John Jermain Porter, "The Manufacture of Pig Iron," in *The ABC of Iron and Steel*, 2d ed., A. O. Backert, ed. (Cleveland: Penton Publishing Co., 1917), pp. 78–80; C. D. King, *Seventy-five Years of Progress in Iron and Steel* (New York: American Institute of Mining and Metallurgical Engineers, 1948), pp. 49–61.
8. The quotation is from the testimony of William Weihe, U.S. Congress, Senate Committee on Education and Labor, *Report Upon the Relations Between Labor and Capital and Testimony Taken by the Committee*, 4 vols. (Washington, D.C.: U.S. Government Printing Office, 1885), 2: 14. See also the testimony of Robert D. Layton, 1: 20–22.
9. Brody, *Steelworkers*, pp. 50–54.
10. Testimony of William McQuade, U.S. Congress, House of Representatives, 52d Cong., 1st Sess., Misc. Doc. 335, *Investigation of the Employment of Pinkerton Detectives in Connection with the Labor Troubles at Homestead, Pa.* (Washington, D.C.: U.S. Government Printing Office, 1897), p. 187.
11. United States, Dept. of Commerce, Bureau of the Census, *Historical Statistics of the United States, Colonial Times to 1970*, 2 Vols. (Washington, D.C.: U.S. Government Printing Office, 1975), 1: 168.
12. Quoted in Wall, *Andrew Carnegie*, p. 345.
13. Brody, *Steelworkers*, pp. 96–100.
14. Charles P. Neill, *Report on Conditions of Employment in the Iron and Steel Industry*, 62d Cong., 1st Sess., Senate Doc. 110, 4 vols. (Washington, D.C., 1911–1913) (hereafter cited as Neill, *Conditions of Employment*), 3: 169–70.
15. John A. Fitch, *The Steel Workers* (New York: Charities Publication Committee, 1911), pp. 11, 201. See also pp. 15, 202–04.
16. Wall, *Andrew Carnegie*, p. 521.
17. For the Carnegie-Jones relationship, see Livesay, *Carnegie and Big Business*, pp. 133–35; Wall, *Andrew Carnegie*, pp. 343–45, 358–59.
18. William Jones, "On The Manufacture of Bessemer Steel and Steel Rails in the United States," *Journal of the Iron & Steel Institute, 1881*, pp. 131–34. Apparently

Jones's address was read for him at the May meeting. He appeared before the institute in person at its meeting of October 11 and discussed the paper.

19. Jones to E. V. McCandless, Feb. 25, 1875, quoted in Bridge, *Inside History of Carnegie Steel*, p. 81.

20. Jones, Address to British Iron & Steel Institute, pp. 131, 139. The British, too, found workmen generally opposed to innovation. See discussion, p. 143.

21. Quoted in Livesay, *Carnegie and Big Business*, p. 133.

22. Jones, Address to British Iron & Steel Institute, pp. 136–37.

23. Andrew Carnegie, *The Gospel of Wealth and Other Timely Essays*, Edward C. Kirkland, ed. (Cambridge, Mass: Belknap Press, 1962), p. 114.

24. The summation that follows is based on the articles by Carnegie, "An Employer's View of the Labor Question," *Forum*, 1 (April 1886): 114–25, and "Results of the Labor Struggle," *Forum*, 1 (August 1886): 538–51, as reprinted in Carnegie, *Gospel of Wealth*, pp. 92–105, 106–22.

25. Carnegie, *Gospel of Wealth*, pp. 115–16. See also, pp. 94–96.

26. Ibid., pp. 101–04.

27. See Robert Hessen, *Steel Titan: The Life of Charles M. Schwab* (New York: Oxford University Press, 1975), p. 35.

28. Carnegie, *Gospel of Wealth*, pp. 113–14, 120.

29. Ibid., pp. 98–99.

30. Ibid., pp. 94, 104.

31. Ibid., pp. 110, 120–22.

32. Wall, *Andrew Carnegie*, p. 522.

33. Ibid., pp. 528–29; Livesay, *Carnegie and Big Business*, pp. 136–37.

34. For brief accounts of the strike, see Livesay, *Carnegie and Big Business*, pp. 139–44, and Wall, *Andrew Carnegie*, pp. 537–82. For a more detailed account, see Leon Wolff, *Lockout: The Story of the Homestead Strike, 1892* (New York: Harper and Row, 1965).

35. Wall, *Andrew Carnegie*, p. 537.

36. Edward W. Bemis, "The Homestead Strike," *Journal of Political Economy*, 2 (June 1894): 378–79; E. R. L. Gould, *The Social Condition of Labor*, Johns Hopkins University Studies in Historical and Political Science, ed. Herbert B. Adams (Baltimore: The Johns Hopkins Press, 1893), pp. 23, 31.

37. Wall, *Andrew Carnegie*, pp. 625–26.

38. Ibid., p. 527. Mahlon M. Garland, who had worked at the Homestead mills, told a congressional investigating committee in 1903 that Carnegie had returned to two shifts after the strike only because he could no longer get enough men to man three shifts. U.S. Congress, Senate, Committee on Education and Labor, 57th Cong., 2d Sess., Senate Doc. 141, *Hearings on Eight Hours For Laborers in Government Contracts* (Washington, 1903), p. 83. This argument presented by Garland, even if true, would not explain why Carnegie continued the two-shift system once workmen became plentiful during the depression of 1893.

39. Dinkey, testifying before the same committee, see pp. 392–95. Neither time period for which he cited statistics can be regarded as typical. The figures for the eight-hour shifts were gathered in the period of uncertainty and hostility immediately preceding the strike. Those for the twelve-hour shifts came in a period of intense pressure on the men when Schwab was eliminating "waste" at Homestead. Jones, in his address to the British Iron & Steel Institute, made a similar argument about men speeding up the longer they worked together. See pp. 130–31.

40. Frank Woodward Miller, "Swissvale Then and Now, A Brief Historical Sketch," *The Twentieth Anniversary of Swissvale Borough* (Swissvale: n.p., 1918), pp. 5, 15.

41. Ibid., p. 15; "Memoirs," Chapter 1 and 4; Dickson, *Paper Read to 28th Annual Reunion and Dinner of Carnegie Veteran Association, December 13, 1929* (n.p., n.d.), copy, box 6, WBD Papers.

42. Bridge, *Inside History of Carnegie Steel*, pp. 150–52.

43. "Memoirs," Chapter 4.

44. Ibid., Family Scrapbook 2.

45. Bridge, *Inside History of Carnegie Steel*, pp. 155–59; Dickson, Address, Elizabeth, N.J., undated, box 3, WBD Papers.

46. On Dickson's career, see "Memoirs," Chapter 4; on courtship and marriage, Chapter 3 and "Love's Young Dream," Family Scrapbook 2, box 2, WBD Papers. The Dicksons had six children: Susan, Emma, Thomas, Eleanor, Charles, and Helen. Thomas, ill with diptheria, died in his father's arms, Dec. 31, 1896.

47. George L. McCague to Dickson, Dec. 20, 1904, box 5, WBD Papers.

48. The biographical data that follows is based on the following sources: For Schwab, Hessen, *Steel Titan*, pp. 3–42. For Corey, *National Cyclopaedia*, 14: 72; Ralph D. Paine, "William Ellis Corey," *World's Work*, 6 (October 1903): 4025–27; *Current Literature*, 35 (December 1903): 700; *Iron Age*, 72 (Aug. 6, 1903): 30. For Dinkey: Hessen, *Steel Titan*, pp. 17–19; *National Cyclopaedia*, 22: 97; *Iron Age*, 72 (Aug. 6, 1903): 30, and 96 (Oct. 7, 1915): 827.

49. Wall, *Andrew Carnegie*, pp. 532–34.

50. Except where indicated, the summary and quotations that follow come from Dickson's "Memoirs," Chapter 4.

51. In his chapter, "The Metamorphic Effects of Power," David Kipnis, *The Powerholders* (Chicago: University of Chicago Press, 1976), pp. 168–211, discusses the psychological impact that the exercise of coercive power over others has on the powerholder's perception of those being commanded. Of course, none of the research cited by Kipnis reproduces the relationships of managers and workers in Carnegie's mills in the 1890s. It does, however, generally support the conclusions that I reached independently.

52. Corey's and Dinkey's views will be documented at length in chapters 5 and 6 below. For Schwab's views, see Hessen, *Steel Titan*, pp. 124, 129, 228, and 229.

53. Wall, *Andrew Carnegie*, p. 799.

54. Appraisal of Dickson's home, 110 Llewellyn Road, Montclair, New Jersey, Jan. 11, 1913, box 9; Family Scrapbooks 1, 3, and 4.

55. The appraisal also included a total for jewelry of $15,170, the two most valuable pieces being a pearl necklace ($6,500) and a diamond pendant ($4,250) belonging to Mrs. Dickson. The four daughters and Mrs. Dickson had furs valued at $1,875, the most valuable being a $500 mink coat.

56. In his diaries, kept in 1907–1912, 1918–1928, and 1932, Dickson listed and commented frequently on his reading. Box 1, WBD Papers.

57. Ibid., see box 3 for Dickson's poetry, pamphlets, addresses, plays, etc. For his published poems, see *Survey*, 26 (June 3, 1911): 344; *The Manufacturers Record*, Feb. 21, 1918.

58. Diaries, box 1, passim.

59. Family Scrapbooks 1 and 3, box 2.

60. Bridge, *Inside History of Carnegie Steel*, pp. 295–96. Bridge notes that the capitalization carried on the books was artificially very low. Carnegie quoted in Burton J. Hendrick *The Life of Andrew Carnegie*, 2 vols. (Garden City, N.Y.: Doubleday, Doran and Co., 1932), 2: 1.

61. Primary accounts of the formation of U.S. Steel will be found in U.S. House of Representatives, Committee on the Investigation of the United States Steel Corporation, *Hearings*, 8 vols, 62d Cong., 2d Sess., 1911–1913 (hereafter cited as Stanley Committee, *Hearings*), vols. 1 and 2. For brief secondary accounts, see Livesay, *Carnegie and Big Business*, pp. 169–89; Wall, *Andrew Carnegie*, pp. 765–93.

62. For the Frick-Carnegie quarrel and split, see note 38, chapter 4 below.

63. Stanley Committee, *Hearings*, 2: 1401. Schwab, testifying at the time, agreed with this characterization of Morgan's role in U.S. Steel.

64. Harold U. Faulkner, "Gary, Elbert Henry," *Dictionary of American Biography*, 7: 175–76.

65. For example, see U.S. Steel Corporation, *Minutes of Annual Meeting of Stockholders on April 15, 1912*, p. 13. Copy, WBD Papers. See also, Ida Tarbell, *The Life of Elbert H. Gary* (New York: D. Appleton, 1925), pp. 20–37.

66. Faulkner, "Gary"; Tarbell, *The Life of Gary*, pp. 76–77.

67. Tarbell, *The Life of Gary*, p. 81.

68. Frederick Lewis Allen, *The Great Pierpont Morgan* (New York: Harper and Brothers, 1949), pp. 164–66; Tarbell, *The Life of Gary*, pp. 83–85.

69. John Kenneth Galbraith, *The Age of Uncertainty* (Boston: Houghton Mifflin, 1977), p. 312.

70. Thomas Cochran, "Gary," *Encyclopedia of American Biography*, ed. John A. Garraty (New York: Harper and Row, 1974), pp. 420–21; Garraty, *Right-Hand Man, The Life of George W. Perkins* (New York: Harper and Brothers, 1957), pp. 94–96; Herbert L. Satterlee, *J. Pierpont Morgan, An Intimate Portrait* (New York: Macmillan, 1939), pp. 345–46.

71. Quoted in Garraty, *Right-Hand Man*, p. 217.

72. Cochran, "Gary," p. 410.

Chapter 2. Promises to Keep

1. Selected minutes of U.S. Steel's board of directors and of committees of the board were published in the Stanley Committee's *Hearings*, pp. 3745–933 (hereafter cited as Minutes). For the strength of the Amalgamated in 1901, see Minutes, Executive Committee, June 17, 1901, p. 3831.

2. Brody, *Steelworkers*, pp. 60–61.

3. The published Minutes did not uniformly indicate who was speaking or how individuals voted. In general, however, it is not difficult to identify the interest of the person speaking. Of the members of the committee, only Steele had no previous experience in the steel industry. A distinguished New York attorney, he had become a partner in J. P. Morgan & Co. in 1899 and a member of U.S. Steel's board in 1901. (*National Cyclopaedia*, 14: 78.) Reid had been a successful banker until 1891 when he founded and became president of the American Tin Plate Co. (*National Cyclopaedia*, 19: 392.) Converse and Edenborn, in the Schwab tradition, had risen through the ranks as steelmen to head, respectively, the National Tube Co. and the American Steel & Wire Co. (See Converse obituary, *New York Times*, April 5, 1921, and "Edenborn," *National Cyclopaedia*, 18: 248–49.) Roberts, by background, fell between the groups. Educated at Haverford, with graduate work in metallurgy and chemistry at the University of Pennsylvania, he rose through the ranks of the Pencoyd Iron Co., owned and founded by his father, and became president of American Bridge (obituary, *New York Times*, March 7, 1943). Roberts's education and inherited wealth seem to have separated him from the self-made steelman. He also disliked Schwab's methods of operation.

4. Tarbell, *The Life of Gary,* p. 156.
5. Minutes, p. 3826.
6. Ibid., pp. 3831–833.
7. Ibid., pp. 3826–827.
8. Ibid., p. 3826.
9. Ibid., pp. 3827–828.
10. Ibid., pp. 3829–831.
11. Ibid., p. 3791.
12. Ibid., pp. 3792–793. For Schwab's reaction, see Hessen, *Steel Titan,* p. 128.
13. Ibid., pp. 3793–794.
14. Ibid., pp. 3832–833, 3819.
15. Ibid., pp. 3820–825.
16. For accounts of the strike, see John A. Garraty, "The United States Steel Corporation Versus Labor: The Early Years," *Labor History,* 1 (Winter 1960): 11–14; Brody, *Steelworkers,* pp. 62–68.
17. Hessen, *Steel Titan,* pp. 132–33.
18. Ibid., pp. 133–36.
19. Tarbell, *The Life of Gary,* p. 149.
20. Hessen, *Steel Titan,* pp. 145–62.
21. Paine, "William Ellis Corey," p. 4025.
22. Garraty, *Right-Hand Man,* pp. 109–14. Perkins is quoted, p. 110.
23. Stanley Committee, *Hearings,* pp. 1574–575.
24. Early draft of memoirs, Family Scrapbook 2, box 2, WBD Papers.
25. *Chicago Tribune,* May 14–16, 1906. I have been unable to determine whether the meeting of the Casualty Managers came before or after the newspaper attack on U.S. Steel.
26. Dickson to Taylor, May 1; Taylor to Dickson, May 20; Dickson to Gary, May 21; Dickson to subsidiary presidents, May 23; Dickson to Gary, June 17, 1907, copies in "Memoirs," Chapter 13, box 2, WBD Papers.
27. Dickson to Gary, June 17, 1907, "Memoirs," Chapter 13.
28. Dickson address to the first session of the American Iron & Steel Institute, Oct. 14, 1910, WBD Papers; Brody, *Steelworkers,* p. 166. I have found that various U.S. Steel officials at the time were giving quite different figures on the number of persons and the amounts of money involved in the safety program. For example, Raynal C. Bolling, assistant general solicitor of U.S. Steel, in an address in April 1911, said that 100 men were involved and that the corporation had spent over $1,000,000 on safety since 1906. See *Annals of the American Academy of Political & Social Sciences* (hereafter cited as AAP&SS, *Annals*) 38: 39.
29. David S. Beyer, "Safety Provisions in the United States Steel Corporation," *Survey,* 24 (May 7, 1910): 205, quoted by Dickson in his address to the American Iron & Steel Institute, Oct. 14, 1910. See also, Daniel Nelson, *Manager and Workers: Origins of the New Factory System in the United States 1880–1920* (Madison: University of Wisconsin Press, 1975), p. 32.
30. MacVeagh to Dickson, Dec. 21, 1909, "Memoirs," Chapter 13; Brody, *Steelworkers,* p. 167.
31. Taylor to Corey, Dec. 2, 1909; Dickson to Taylor, Mar. 2, 1910, "Memoirs," Chapter 12.
32. Charles A. Gulick, *Labor Policy of the United States Steel Corporation* (New York: Columbia University Press, 1924), p. 156.
33. Raynal C. Bolling, "Rendering Labor Safe in Mine and Mill," American Iron &

Steel Institute (hereafter cited as AI&SI), *Yearbook, 1912*, pp. 106–13. The quotation is from p. 111.

34. Brody, *Steelworkers*, pp. 86–89; David Brody, *Labor in Crisis: The Steel Strike of 1919* (Philadelphia: J. B. Lippincott, 1965), pp. 89–95.

35. Gulick, *Labor Policy of U.S. Steel*, pp. 180–84. I have calculated the percentages from the statistical charts provided by Gulick, pp. 180–83.

36. "Memoirs," Chapter 10.

37. Draft resolution dated April 6, 1907. For the wording of the resolution as adopted on April 23, see Dickson to W. B. Palmer, April 26, 1907, box 7, WBD Papers. Also, see Dickson to Corey, Apr. 16, 1907, "Memoirs," Chapter 7.

38. Neill, *Conditions of Employment*, 3: 165–66. Dickson to Corey, April 16, 1907, "Memoirs," Chapter 7.

39. For a recent study of the Pittsburgh Survey, see John T. McClymer, "The Pittsburgh Survey, 1907–1914: Forging an Ideology in the Steel District," *Pennsylvania History*, 41 (April 1974): 169–86. Also, see Dickson to E. M. Hagar, Feb. 23, 1909; Dickson to Corey, March 15, 1910, "Memoirs," Chapter 7, WBD Papers. Dickson did not keep a copy of his report nor does he state which mills worked on Sunday or what overall number or percentage of men worked seven days a week. In his study for the Pittsburgh Survey, John A. Fitch estimated that at least twenty-five percent of all steelworkers in Allegheny County, Pennsylvania, where several of U.S. Steel's largest plants were located, worked the seven-day workweek. Fitch, *Steel Workers*, p. 175.

40. Hagar to Dickson, March 1, 1909, "Memoirs," Chapter 7.

41. Ibid., Dickson to Corey, March 2, 1909.

42. Ibid., Dickson to Corey, May 10, 1909. Dickson may have been wrong on this point. Despite Carnegie's dominant position in the industry, his lead in adopting the eight-hour shift in the 1880s was not followed by the rest of the industry. See above, pp. 10 and 13–15.

43. Dickson to Corey, Jan. 26, 1910, "Memoirs," Chapter 7.

44. Ibid. The summation of the meeting and the quotations are all from an extract copy "From Minutes of Presidents' Meeting, Feb. 25th, 1910."

45. Ibid., Dickson to Samuel Conover, et al, April 12, 1935.

46. Ibid., Dickson to Dinkey, March 3; Dinkey to Dickson, March 14; both quoted in Dickson to Corey, March 15, 1910 (italics added).

47. Ibid., Dickson to Corey, March 17, 1910.

48. Kellogg's address to Silver Anniversary of Survey Associates, Inc., Dec. 2, 1937, copy, box 5, WBD Papers.

49. For accounts of the Bethlehem Strike at the time, see *Survey*, 24 (May 21, 1910): 306–08; (July 2, 1910): 576–78. For the beginning of the investigation, see *Washington Post* and *New York Times*, March 16, 1910. On April 19, Congress directed that the study be reported to it. *Congressional Record*, 45: 4957. For an excellent recent study of the strike, see Robert Hessen, "The Bethlehem Steel Strike of 1910," *Labor History*, 15 (Winter 1974): 3–18.

50. Dickson to Corey, March 15, 1910, "Memoirs," Chapter 7.

51. Ibid., Dickson's comments regarding letter to Corey, March 15, 1910.

52. Ibid.; *Congressional Record*, 45: 3417–418.

53. "Memoirs," Chapter 7.

54. Ibid., draft, E. H. Gary to Presidents of U.S. Steel subsidiaries, dated March 18, 1910.

55. Kellogg address, Dec. 2, 1937, box 5, WBD Papers; Gulick, *Labor Policy of U.S. Steel*, pp. 25–30.

56. Dickson to Gary, May 12, 1910, box 7, WBD Papers.

Chapter 3. Next to President of the United States . . .

1. Copy, Dickson to Samuel Conover, et al., Apr. 12, 1935, "Memoirs," Chapter 2, WBD Papers.
2. "Memoirs," Chapter 9.
3. *New York Journal*, *New York Times*, *New York Tribune*, Mar. 17, 1911, and *New York Evening Post*, Mar. 18, 1911. The resignation of Thomas Morrison, a Carnegie partner, from U.S. Steel's board of directors to make way for Farrell lent support to the rumor. On the other hand, among the high-ranking Carnegie men still at U.S. Steel were Second Vice-President David G. Kerr, who would remain till his retirement, and Alva C. Dinkey, who continued as president of Carnegie Steel until 1915.
4. Copy, Dickson to Conover, et al., Apr. 12, 1935, "Memoirs," Chapter 2.
5. Concurring opinion, Judge Victor B. Wooley, *U.S.* vs. *U.S. Steel Corp.*, 223 Federal Reporter 173–74 (June 3, 1915).
6. "Memoirs," Chapter 9.
7. Dickson to Corey, Aug. 10, 1904, box 7, WBD Papers.
8. Kirby Page, "Gold from Steel," *The World Tomorrow*, Apr. 12, 1933, p. 248, gives total earnings, dividends, undivided surplus, and wages and salaries for U.S. Steel between 1901 and 1932, based on annual reports of the corporation.
9. Gary statement, Feb. 18, 1909, quoted in *Iron Age*, 83 (Feb. 25, 1909): 648. U.S. Steel's position at the onset of the panic was good, Gary told the directors on Oct. 29, 1907. "As a rule, the stocks on hand of our customers, jobbers and others, are unusually low. . . . Our inventories are not large, the total stock of pig iron, for instance," was "about the medium in the history of the Corporation." *Iron Age*, 80 (Oct. 31, 1907): 1236–237.
10. Notes, "Elbert H. Gary," box 6; copy, Dickson to Corey, Feb. 16, 1909, box 7, WBD Papers; *Iron Age*, 80 (1907): 1245, 1633.
11. Copy, Dickson to Morgan, June 6, 1935; "Memoirs," Chapter 16.
12. Notes, "Gary," box 6, WBD Papers.
13. Stanley Committee, *Hearings*, pp. 75–78.
14. Tarbell, *The Life of Gary*, p. 205.
15. Ibid., p. 206.
16. *Iron Age*, 80 (Nov. 28, 1907): 1549, reports the views of Gary's guests.
17. Ibid., 84 (Sep. 9, 1909): 786–87.
18. Stanley Committee, *Hearings*, p. 280.
19. *Iron Age*, 81 (May 28, 1908): 1710.
20. Quoted, Garraty, "U.S. Steel vs. Labor," p. 24.
21. Quoted, Tarbell, *The Life of Gary*, p. 205.
22. Dickson to Corey, Feb. 16, 1909, box 7, WBD Papers. Unfortunately, Dickson supplied no statistics to support any of his contentions.
23. *New York Times*, Feb. 20, 1909. See also, *Iron Age*, Feb. 18, 1909.
24. Quoted, Garraty, "U.S. Steel vs. Labor," p. 24.
25. Quoted, Brody, *Steelworkers*, p. 151; *Iron Age*, 83 (Apr. 8, 1909): 1136.
26. The figures cited for 1907 and 1908 are from Gary's statement of Feb. 18, 1909, quoted in *Iron Age*, Feb. 25, 1909. The figure of "over 30,000" tons is from Garraty, "U.S. Steel vs. Labor," p. 25. Garraty, however, refers to this as "a healthy increase over previous months." Industry-wide production of pig iron reached a low daily average tonnage of under 34,000 in Jan. 1908. Output was irregular until June 1908, after which it

climbed steadily into 1909. According to Neill, *Conditions of Employment*, 3: 208, the average number of men employed in blast furnace work followed a similar industry-wide pattern. A low of 18,500 men was reached in Jan. 1908. After irregular rises and falls, a steady gain began in June 1908 and continued upward (except for March and April 1909) until January 1910.

27. Garraty, "U.S. Steel vs. Labor," p. 25; *Iron Age*, 83 (June 17, 1909): 1929. For industry-wide wage rates, month by month, see Neill, *Conditions of Employment*, 3: 252.

28. Gulick, *Labor Policy of U.S. Steel*, Table V, p. 57.

29. Garraty, "U.S. Steel vs. Labor," p. 25.

30. Newsclipping, *Inter-Ocean*, Apr. 2, 1909, box 6, WBD Papers.

31. Garraty, "U.S. Steel vs. Labor," pp. 25–26.

32. *Wall Street Journal*, Apr. 2, May 5–6, 1909.

33. *The Annual Statistical Report of the American Iron & Steel Institute for 1914* (Philadelphia, 1915), p. 76, shows U.S. Steel's portion of pig iron, steel ingots and castings, and finished rolled products by year from 1901 through 1914. For the period in question, the following are reported:

Year	*Pig Iron*	*Steel Ingots*	*Rolled Steel*
1907	44.3 %	57.1 %	48.5 %
1908	43.5 %	55.9 %	47.1 %
1909	45.0 %	55.8 %	48.1 %
1910	43.3 %	54.3 %	48.1 %

For detailed price changes on a variety of steel products by month, see *Iron Age*, 85 (Jan. 6, 1910): 27.

34. T. J. Drummond, president, Lake Superior Corporation, *Iron Age*, 84 (Oct. 21, 1909): 1217. *Iron Age* covered the dinner in detail, pp. 1217–222, and included a list of all the guests.

35. Ibid., p. 1218.

36. Ibid., p. 1219.

37. Ibid., p. 1220.

38. Ibid., p. 1221, italics added.

39. Memorandum of meetings, Jan. 28 and Nov. 30, 1910, box 7, WBD Papers.

40. Garraty, *Right-Hand Man*, pp. 120–22.

41. Copies, Gary to Corey, Corey to Gary, Dec. 31, 1909; Gary to Corey, Jan. 2, 1910, box 7, WBD Papers.

42. Garraty, *Right-Hand Man*, pp. 122–23; copy of the petition with signatures, dated Jan. 3, 1910, box 7, WBD Papers.

43. Undated memo, box 7, WBD Papers.

44. Garraty, *Right-Hand Man*, pp. 123–25. According to Garraty, the resolution regarding the authority of the chairman was adopted May 1. He cites Tarbell, *The Life of Gary*, p. 219, however, and Tarbell gives the date as March 1.

45. See resignation and letters to Corey, Jan. 5, 1910, with notations, box 7, WBD Papers. The letter to Corey is dated Jan. 5, 1909, but I doubt that Dickson resigned on both Jan. 5, 1909 and Jan. 5, 1910. It being the first of a new year, he probably misdated the letter. Dickson had turned in a resignation once before, on Apr. 11, 1905, but Corey had refused to accept it.

46. Introduction to "Memoirs"; Dickson to Tom M. Girdler, June 4, 1937, "Memoirs," Chapter 8.

47. *Iron Age*, 85 (June 2, 1910): 1273–274, reports Dickson's address and the commentary in full.

48. Ibid., pp. 1274–275. The Gary quotation is indirect.

49. "Memoirs," Chapter 8. For comment on the address, see *Iron Trade Review*, June 9, 1910; *Survey*, 24 (June 18, 1910): 475–77.

50. Stanley Committee, *Hearings*, pp. 81–82.

51. *Address of the President and Papers Delivered at the American Iron & Steel Institute, First Formal Meeting, Waldorf-Astoria, New York City, Oct. 14, 1910*, pp. 4–10. Farrell's remarks, pp. 11–19. A copy can be found in WBD Papers.

52. Ibid., pp. 20–31.

53. "Memoirs," Chapter 8.

54. *Iron Age*, 86 (Oct. 30, 1910): 1030, for Dickson's address. For the commentaries of Kirchhoff, Cook, and Bailey, see 86 (Nov. 24, 1910): 1220–221.

55. See comments in *Iron Trade Review*, Nov. 3, 1910; *Manufacturers Record*, undated clipping, "Memoirs," Chapter 9.

56. See *New York Times*, Dec. 1905–Jan. 1906; Aug. 1–3, 1906; May 1907. See also, Garraty, "U.S. Steel vs. Labor," p. 27.

57. Undated, unidentified newsclipping, 1906, box 7, WBD Papers.

58. *New York Times*, Feb. 5, 1910.

59. Garraty, *Right-Hand Man*, p. 125; memorandum of meeting, Nov. 10, 1910, box 7, WBD Papers.

60. Garraty, *Right-Hand Man*, p. 125; *Iron Age*, 87 (Jan. 5, 1911): 7.

61. *New York Times*, Jan. 11, 1911.

62. Ibid.; see also, *New York Tribune*, *Wall Street Journal*, Jan. 11, 1911.

63. *New York Times*, Jan. 11, 1911.

64. U.S. Steel folder 2, box 7, WBD Papers.

65. Ibid., memorandum, Jan. 16, 1911.

66. Ibid., memorandum, Feb. 6, 1911; telegram, Gary to Dickson, Feb. 9, 1911.

67. See *New York Times*, *New York Journal*, Mar. 17, 1911; *New York Evening Post*, Mar. 18, 1911; *Iron Trade Review*, Mar. 23, 1911. For the rumors regarding the new steel trust, see unidentified clipping, "Dickson Quits the Trust," box 7, WBD Papers.

68. *New York Times*, Mar. 17, 1911.

Chapter 4. The Battle Continues

1. *Satires and Epistles of Horace*, trans. S. P. Bovie (Chicago: University of Chicago Press, 1959), II.6, p. 138.

2. Diary, box 1, WBD Papers. Dickson wrote in his diary almost daily between May 1911 and January 1912. Entries during 1912 were intermittent and he apparently kept no diary between 1913 and 1918. His retirement ended in the autumn of 1915.

3. Early draft of memoirs, Family Scrapbook 2, box 2, WBD Papers.

4. Diary entries, July 31, Aug. 5, 1911; Oct. 6 and 8, 1912.

5. Subpeona, dated May 23, 1911; Dickson's comments, U.S. Steel folder 2, box 7, WBD Papers.

6. Diary entries, May 25, 29, 21; June 1 and 2, 1911.

7. Henry E. Colton to Dickson, Nov. 8; Dickson to Colton, Nov. 11, 1912, U.S. Steel folder 3, box 7, WBD Papers.

8. Ibid., undated memo, written in 1912.

9. "Memoirs," Chapter 19, box 2, WBD Papers. Roy Lubove, *The Struggle for Social Security, 1900–1935* (Cambridge, Mass.: Harvard University Press, 1968), p. 55, says,

"The National Association of Manufacturers was a leading advocate of workmen's compensation." That was true only after the organization's annual convention in May 1911, following adoption of the New Jersey law. Prior to that date state associations apparently acted on their own.

Authorized to make a thorough study of the question of employer liability, NAM staff members, led by F. C. Schwedtman and James A. Emery, in 1910 surveyed 25,000 American businessmen and examined the employer liability and accident compensation systems of other nations. In 1911 the NAM published their findings under the title *Accident Prevention and Relief,* and distributed the book at the NAM convention in May. Scholars frequently cite and sometimes quote from an address that Schwedtman made before the American Academy of Political and Social Sciences (AAP&SS, *Annals,* 38: 202–04) to prove the enthusiasm of businessmen for employer liability laws. In his remarks, Schwedtman declared the survey of 25,000 American businessmen indicated that "more than ninety-five per cent of those answering were in favor of an equitable, automatic compensation system for injured workers and their dependents." This statement must be used with caution. In *Accident Prevention and Relief,* p. xiii, Schwedtman pointed out that only 10,000 replies were received to the 25,000 inquiries mailed out. Later on in his address to the AAP&SS, Schwedtman somewhat watered down his initial statement: "Thousands of interviews and letters convince me that many employers would welcome compulsory legislation on the subject, if only somebody would tell them how it is to be accomplished in a manner which is constitutional as well as just to all concerned." He noted that "many small employers" lacked the means or ability to launch voluntary programs and could not afford systems "outlined by the representatives of large and wealthy concerns." Clearly there is something of a gap between acceptance of an abstract principle and endorsement of a specific remedy.

10. AAP&SS, *Annals,* 38 (Apr. 8, 1911): 218–20.

11. Ibid., pp. 223–24.

12. Dickson to Fielder, Feb. 4, 1914, copy in "Memoirs," Chapter 19.

13. Lubove, *Struggle for Social Security,* pp. 58–59.

14. Kellogg to Dickson, Mar. 16, 1911, "Memoirs," Chapter 19.

15. Copy, Kellogg to Bertram, June 20, 1911, U.S. Steel folder 2, box 7, WBD Papers.

16. Diary entry for Oct. 17, 1911.

17. Copies, Kellogg to Page, Oct. 17; Kellogg to Johnson, Oct. 26, 1911, U.S. Steel folder 6, box 7, WBD Papers.

18. "Charles M. Cabot—Industrial Reformer," *Survey,* 35 (Oct. 2, 1915): 27–29; Frank Barkley Copley, "A Great Corporation Investigates Itself," *American Magazine,* 74 (Oct. 1912): 643–49. Charles Hill, "Fighting the Twelve-Hour Day in the American Steel Industry," *Labor History,* 15 (Winter 1974): 19–35, gives a good account of the role of the reformers in ending the twelve-hour day in the steel industry.

19. Copley, "A Great Corporation Investigates Itself," p. 648.

20. "Memoirs," Chapter 10.

21. "Cabot," *Survey,* 35: 28; Cabot to Stockholders, Mar. 26, 1912, copy, "Memoirs," Chapter 10.

22. Gary to Stockholders, Mar. 12, 1912, copy, box 7, WBD Papers.

23. U.S. Steel Corporation, *Minutes of Annual Meeting of Stockholders, Apr. 15, 1912,* pp. 1–2. Copy in box 7, WBD Papers.

24. *Report of Committee of Stockholders of the United States Steel Corporation, Apr. 15, 1912,* pp. 4–5. Copy, box 7, WBD Papers.

25. Ibid., pp. 5–7.

26. Ibid., pp. 7–8.
27. Ibid., p. 8.
28. Ibid., pp. 9–17.
29. U.S. Steel Corp., *Minutes, Annual Meeting, 1912*, pp. 5–7, gives Gary's analysis of the letters. For the quotations, see *Copies of Letters Received from Stockholders in Answer to the C. M. Cabot Circular Letter of March 26, 1912*, pp. 2, 5, 6, 12, 29, 30, 34, 59, 62, 63, and 68. Copy, Gary folder, box 7, WBD Papers.
30. U.S. Steel Corp., *Minutes, Annual Meeting, 1912*, p. 6.
31. Ibid., pp. 7–8.
32. Ibid., pp. 9–11.
33. Ibid., pp. 11–13.
34. Gary to stockholders of U.S. Steel Corp., May 28, 1912, printed under title, *Action of United States Steel Corporation Upon Recommendations of Stockholders' Committee*. Copy, Gary folder.
35. Fitch, "Holding Fast to the Twelve Hour Day," 165–66.
36. Bull, "The Twelve-Hour Shift in the Steel Foundry," *Iron Age*, 90 (Oct. 3, 1912): 808–09.
37. *Survey*, 31 (Jan. 3, 1914): 376.
38. Deeply embarrassed at the unfavorable publicity generated by the Homestead Strike, Carnegie afterward minimized his role and hinted that the trouble was caused by Frick's mismanagement of the affair. The two men had agreed in advance, however, that the union must be destroyed and Carnegie gave Frick a free hand to manage the strike as he thought best. See Wall, *Andrew Carnegie*, pp. 537–82. The final break grew out of an attempt by a syndicate to purchase Carnegie Steel in 1899. Carnegie was anxious to sell, and Frick carried on the negotiations with the syndicate whose members remained anonymous. Carnegie insisted that a deposit of $2 million be made on the option to buy. Frick put up part of the money. When the group failed to raise the money needed to complete the transaction, the deal fell through. Carnegie pocketed $1.7 million—his share of the deposit—which included the money contributed by Frick. Carnegie was angered when he discovered that the syndicate was made up of notorious speculators with whom he would have been ashamed to deal and that Frick was to receive a secret bonus of $2.5 million for helping to complete the transaction. Troubles between Frick and Carnegie multiplied once the Old Scot lost faith in his partner, and eventually Carnegie decided to buy out Frick under the terms of the company's "iron clad agreement." Whenever a junior partner died, left the company, or was voted out by at least two-thirds of the partners, he or his estate was paid book value rather than actual value for his interest. Originally book and actual value had not been far apart, but by 1899 book value represented only a fraction of market value. Frick resisted and brought suit in the courts. A stalemate followed and eventually the matter was compromised. Carnegie Steel incorporated with a capital of $320 million, Frick was removed from all active participation in the company, and he received over $30 million for his interest instead of the $4.9 million due him under the iron clad agreement. See Wall, *Andrew Carnegie*, pp. 723–63.
39. Early draft of memoirs, Family Scrapbook 2.
40. Ibid., Dickson to Carnegie, Apr. 2, 1914, Andrew Carnegie folder, box 6.
41. Ibid., Dickson's comments appear at the end of the letter.
42. Carnegie, among other things, said in "An Employer's View of the Labor Question," *Forum*, 1 (April 1886): 114–25, "I, therefore, recognize in trades-unions, or, better still, in organizations of the men of each establishment, who select representatives to speak for them, a means, not of further embittering the relations between employer and employed, but of improving them."

43. U.S. Steel Corp., *Minutes, Annual Meeting, 1914*, pp. 6–7. Copy, WBD Papers. Although U.S. Steel's earnings in 1913 (over $137 million) were the highest since 1907 (nearly $161 million), 1914 earnings proved to be the lowest yet experienced by the firm ($71 million).

44. Ibid., pp. 7–8.

45. Ibid., p. 9.

46. Ibid., pp. 24–25.

47. Ibid., pp. 26–28.

48. Ibid., p. 29.

49. Ibid., pp. 15–16.

50. Ibid., p. 29.

51. Ibid., pp. 2–9.

52. Brody, *Steelworkers*, pp. 164–65.

53. Brody, *Labor in Crisis*, pp. 56–60, 71.

54. Gulick, *Labor Policy of U.S. Steel*, pp. 29, 40.

55. Ibid., Gary quote, p. 36.

56. Ibid., p. 37. See also, Brody, *Steelworkers*, p. 197.

57. U.S. Steel Corp., *Minutes, Annual Meeting, 1912, 1913, 1914* Gary to stockholders of U.S. Steel Corp., Mar. 12, 1912, box 7, WBD Papers.

58. U.S., Dept. of Commerce, Bureau of the Census, *Historical Statistics of U.S., Colonial Times to 1970*, 1: 172–73, shows the average daily hours for all steelworkers in 1911 as 10.39. All other occupations listed ranged between 9.08 and 9.70 hours per day. In 1901 comparable figures were 10.66 for steel and a range of 8.94 to 9.81 for other industries.

59. Gulick, *Labor Policy of U.S. Steel*, pp. 57, 182.

60. Raynal C. Bolling, "Rendering Labor Safe in Mine and Mill," American Iron & Steel Association, *Yearbook, 1912*, pp. 106–13. Bolling noted that since adoption of the accident relief program only two-tenths of 1 percent of the men injured brought suits against the corporation.

61. Committee of Stockholders, *Report, 1912*, pp. 5–6; Gulick, *Labor Policy of U.S. Steel*, p. 57. The report of the congressional committee that investigated the steel industry in the wake of the Bethlehem Strike of 1910 concluded that a 6 percent increase in costs would be required to enable all continuous operations in the entire industry to move to the eight-hour shift without any loss in wages by the men. That calculation also involved raising the pay of all steel employees who worked more than sixty hours per week. See Neill, *Conditions of Employment*, 3: 175–81.

62. Tarbell, *The Life of Gary*, pp. 91 and 111; *Iron Age*, 88 (June 8, 1911): 1405; 89 (Jan. 18, 1912): 210. Gary's selection of Farrell as president of U.S. Steel may have indicated the importance of the export trade, as Farrell had previously headed the subsidiary that handled all of U.S. Steel's overseas exports.

63. "Memoirs," Chapter 14; early draft of memoirs, Family Scrapbook 2; *Wall Street Journal*, Oct. 6 and 11, 1915; *New York Times*, Oct. 7, 1915.

64. *Wall Street Journal*, Oct. 11, 1915.

65. *Iron Age*, 96 (Sep. 30, 1915): 778.

66. All quotations are from *Wall Street Journal*, Oct. 11, 1915. See also, *Iron Age*, 96 (Oct. 14, 1915): 908–09; *New York Times*, Oct. 9, 1915.

67. *New York Times*, Oct. 9, 1915.

68. *New York Evening Post*, Oct. 9, 1915; *Wall Street Journal*, Oct. 12 and 21, 1915.

69. *Literary Digest*, 51 (Oct. 30, 1915): 947–48.

70. *Wall Street Journal*, Oct. 11, 12, 14, and 21, 1915.

71. Ibid., Oct. 11, 14, and Dec. 4, 1915.

72. Early draft of memoirs, Family Scrapbook 2; *New York Times,* Feb. 8, 1916. For output figures, see Corey's statement, *New York Times,* Feb. 19, 1916. For capacity figures, see *Nation,* 102 (Feb. 17, 1916): 204–05.

73. *Wall Street Journal,* Oct. 23, 1915.

74. *Iron Age,* 98 (Oct. 5, 1916): 787.

75. Based on annual report for 1918, summarized in *Iron Age,* 101 (Apr. 4, 1919): 914.

76. Dickson to Baker, Apr. 13, 1917, co-signed by John A. Topping, chairman, Republic Steel, and E. A. S. Clarke, president, Lackawanna Steel. Copy, "Memoirs," Chapter 14.

77. "Memoirs," Chapter 14.

78. Dickson to editor, *Iron Age,* Aug. 2, 1917, copy in "Memoirs," Chapter 14. On Feb. 8, 1918, Dickson recorded in his diary that "our Wilmington plant" had been closed indefinitely because of high costs of operation. "The prices fixed by the Govt. not affording any margin of profit. Another instance of the futility of interference with natural laws of trade."

79. Dickson, "Some War Problems in the Steel Trade," *Iron Age,* 102 (Oct. 24, 1918): 1015–016.

80. Dickson, "War Tax and Bond Issues," ibid., 101 (June 20, 1918): 1610–611.

81. Dickson to editor, *Iron Age,* Aug. 2, 1917, copy in "Memoirs," Chapter 14.

82. Diary entry for Mar. 14, 1919. Dickson paid an income tax of $19,444. "A heavy price to pay for crushing out autocracy but not too heavy if it is only a permanent job."

83. Diary entry, Dec. 22, 1918.

84. Ibid., Nov. 11, 1918.

85. Ibid., Feb. 11, 1918.

86. Early draft of memoirs, Family Scrapbook 2.

Chapter 5. Experiment with "Industrial Democracy"

1. Diary entry, Sep. 19, 1918, box 1, WBD Papers.

2. Diary entries, Sep. 21 and 22, 1918; "Memoirs," Chapter 10, Box 2, WBD Papers. A copy of the notice posted in the shops is included.

3. "Memoirs," Chapter 10. A copy of Dickson's statement that was read in the plants is included.

4. Dickson's opening remarks, "Memoirs," Chapter 10.

5. For the published minutes of the plant committees, see, *Plan of Representation of the Employees of Midvale Steel & Ordnance Company, Cambria Steel Company and Subsidiary Companies,* pp. 14–20. Copy in Midvale folder 5, box 7, WBD Papers.

6. Family Scrapbook 2, box 2, WBD Papers.

7. See above, pp. 17–18. See also, Family Scrapbooks 2 and 4.

8. Family Scrapbook 4; "Memoirs," Chapter 5.

9. Address, "Eight-Hour Day and Six-Day Week in the Continuous Industries," delivered before the American Association of Labor Legislation, Columbus, Ohio, Dec. 29, 1916 and published in *American Labor Legislation Review,* 7 (Mar. 1917). Sometime earlier in 1916, Dickson used the same words in an address at a church in Elizabeth, N.J. WBD Papers.

10. Ibid.

11. Dickson to Carnegie, Apr. 2, 1914, Andrew Carnegie folder, box 6; copy of address to Friendly Sons of St. Patrick, Jan. 20, 1915, box 3, WBD Papers.

12. Melvin I. Urofsky, *Big Steel and the Wilson Administration* (Columbus: Ohio State University Press, 1969), pp. 192–247; Robert D. Cuff, *The War Industries Board:*

Business-Government Relations During World War I (Baltimore: The Johns Hopkins University Press, 1973), pp. 122–31.

13. Philip Taft, *Organized Labor in American History* (New York: Harper and Row, 1964), pp. 310–11; Urofsky, *Big Steel and Wilson Administration*, pp. 248–61; Brody, *Steelworkers*, pp. 202–08.

14. U.S., Dept. of Labor, Bureau of Labor Statistics, Bulletin #287, *National War Labor Board* (Washington, 1922), pp. 28–34.

15. Carroll E. French, *The Shop Committee in the United States*, Johns Hopkins University Studies in History and Political Science, vol. 41, #2 (Baltimore: The Johns Hopkins Press, 1923), pp. 17–26 (quote, p. 25); National Industrial Conference Board, Research Report #21, *Works Councils in the United States*. . . . (Boston, 1919), p. 13.

16. Diary entry, Feb. 14, 1918.

17. For Bethlehem's labor problems, see Urofsky, *Big Steel and Wilson Administration*, pp. 266–69.

18. National War Labor Board, Case file #129, Employees vs. Midvale Steel & Ordnance Co., National Archives, Record Group 2, Federal Records Center, Suitland, Maryland (hereafter cited as NWLB, Midvale Case file), Transcript of Hearings, pp. 16, 421.

19. Ibid., copy, Kelton to Midvale Steel Co., Apr. 30, 1918, pp. 512–13.

20. Resume, Midvale Case; Kelton to NWLB, July 15, 1918, NWLB, Midvale Case file.

21. Diary entry, Aug. 7, 1918, (memo attached). Dickson did not indicate whether or not the employees could form an organization with ties to a regular union.

22. Diary entry, Aug. 23, 1918: "Drafted a letter for ACD to sign addressed to J. P. Perkins of War Labor Board."

23. Report, Perkins to NWLB, Aug. 28, 1918, Midvale Case file.

24. Urofsky, *Big Steel and Wilson Administration*, pp. 271–78; Brody, *Steelworkers*, pp. 208–13.

25. Brody, *Steelworkers*, pp. 212–13.

26. Telegram, Daniels to Midvale Steel Co., Sep. 13, 1918, copy in "Memoirs," Chapter 10; diary entry, Sep. 16, 1918.

27. Diary entry, Sep. 16, 1918.

28. "Memoirs," Chapter 10.

29. Ibid.; diary entry, Sep. 19, 1918; Dickson's notes, Midvale folder 5, box 7, WBD Papers; Dickson to Samuel Conover, et al, Apr. 12, 1935; Corey to Daniels, Sep. 19, 1918, "Memoirs," Chapter 10.

30. "Memoirs," Chapter 10; diary entries for Sep. 21 and 22, 1918.

31. *Plan of Representation of the Employees of the Midvale Steel and Ordnance Company* . . . , Midvale folder 5, box 7, WBD Papers.

32. "Memoirs," Chapter 10.

33. *Iron Age*, 102 (Sep. 26, 1918): 764; John Calder, "Five Years of Employee Representation under 'The Bethlehem Plan,' " ibid., 111 (June 14, 1923): 1689–696.

34. Compare Midvale Plan, WBD Papers, with Colorado Fuel & Iron Co. Plan, Ben M. Selekman and Mary Van Kleeck, *Employees' Representation in Coal Mines, A Study of the Industrial Representation Plan of the Colorado Fuel & Iron Co.* (New York: Russell Sage Foundation, 1924), pp. 401–37.

35. NWLB, Midvale Case file, Transcript of Hearings, pp. 3–7.

36. Ibid., testimony, IAM Business Manager, Wm. A. Kelton, p. 25.

37. Ibid., Kelton testimony, pp. 25–26; testimony, Edmund J. Cotton, elected representative and IAM member, pp. 42–46, 52, 63, and 72–73.

38. Ibid., testimony, Dickson, pp. 399 and 404.
39. Ibid., pp. 414–23.
40. *The Steelworker* (Midvale Company magazine), Jan. 1919, pp. 5, 7, and 9, published Dickson's and Dinkey's address and the minutes of the session at Philadelphia. Copy, Midvale folder 5, box 7, WBD Papers.
41. Diary entry, Nov. 23, 1918.
42. A copy of the magazine with a note on the reaction of his superiors, Midvale folder 5.
43. U.S., Dept. of Labor, Bureau of Labor Statistics, *NWLB*, p. 182.
44. O'Brien to E. B. Woods, chief administrator, NWLB, Mar. 4, 1919, NWLB, Midvale Case file.
45. Ibid., O'Brien to Woods, Mar. 11, 12, 14, and 22, 1919.
46. Copy, Midvale folder 5. That the document is O'Brien's work I determined from parallel wording in O'Brien's letters to Woods.
47. Unidentified clipping, Midvale folder 5.

Chapter 6. Test, Failure, and Collapse

1. *Johnstown Tribune*, Aug. 25, 1919; diary entry, Aug. 22 and 23, 1919, box 1, WBD Papers.
2. *Wall Street Journal*, Aug. 26, 1919; *Washington Post*, *Philadelphia Inquirer*, and *Pittsburgh Post*, Aug. 25, 1919.
3. A copy of the resolution will be found in Midvale folder 5, box 7, WBD Papers.
4. Ibid. Although none of the newspaper items used the entire release, the portions that were used were verbatim or closely paraphrased versions.
5. *Amalgamated Journal*, 21 (Sep. 18, 1919): 8; Brody, *Steelworkers*, p. 235.
6. For the complete resolution, see Midvale folder 5.
7. Brody, *Steelworkers*, pp. 229–30. The personnel of the NWLB changed with increasing frequency after the Armistice. See *NWLB*, p. 12.
8. Diary entry, Mar. 1, 1919.
9. Copy of address, folder 1, box 3; Dickson to L. C. Fogg, May 14, 1938, folder 2, box 3; diary entry, Feb. 18, 1919, WBD Papers.
10. Foster testimony before Interchurch World Movement Committee, box 26, David A. Saposs Labor Collection, State Historical Society of Wisconsin. I am indebted to Professor Robert Asher for supplying me with copies of data relevant to the strike at Johnstown from this collection. His article, "Painful Memoies: The Historical Consciousness of Steel Workers and the Steel Strike of 1919," *Pennsylvania History*, 45 (January 1978): 61–86, explores the impact that workers' memories of the Homestead Strike and the steel strikes of 1901 and 1909 had on their attitudes during the 1919 strike.
11. Saposs, "Organizing the Steel Workers," Saposs Labor Collection, State Historical Society of Wisconsin, p. 20. Saposs based his report on the minutes of the National Committee. (Microfilmed copy, Historical Collection and Labor Archives, The Pennsylvania State University Library.)
12. George Soule, typewritten ms., "History of the Strike in Johnstown," prepared for the Interchurch World Movement's Commission of Inquiry, Heber Blankenhorn Papers, Archives of Industrial Society, University of Pittsburgh. Copy in Labor Archives, The Pennsylvania State University Library.
13. Diary entry, Jan. 8, 1919. Notes on 1919 Strike, Family Scrapbook 2, box 2, WBD Papers. See also, Saposs, "Organizing the Steel Workers," pp. 22–24.
14. Diary entries, Mar. 16–18, 1919; Notes on 1919 Strike, Family Scrapbook 2.

15. Diary entries, Mar. 21–24, 1919; Dickson to A. A. ("Fred") Corey, Jan. 19, 1920, Midvale folder 7, box 5.

16. The text of Corey's letters of Apr. 7 and May 7, 1919, are given in Saposs, "Organizing the Steel Workers," pp. 76–78. Soule, "History of the Strike in Johnstown," quotes only the first part of Corey's letter of Apr. 7 and then charges him with going back on his implied pledge to work with the union.

17. "Some Twentieth Century Problems," *AAP&SS*, 85 (Sep. 1919); *System*, 35 (June 1919): 1041–044; diary entries, Mar. 1 and 5, 1919.

18. *Iron Age*, 104 (July 17, 1919): 182–83, gives details of the pension and home-building loan plans of Midvale; diary entries, May 10, 12, and June 2–7, 1919.

19. *Johnstown Tribune*, May 12, 1919; diary entries, May 10 and 12, 1919.

20. *Iron Age*, 104: 182–83.

21. *Johnstown Tribune*, May 12, 1919.

22. Diary entries, Sep. 3 and 11, 1919.

23. Dickson to Chadbourne, Sep. 27, 1919. Three days later Dickson sent Chadbourne a copy of Foster's book on syndicalism, which he had read. Midvale folder 5. Dickson wrote in his diary on Sep. 10: "My sympathies are on the side of the Corporation not because I approve of their methods but because of the anarchistic and irresponsible character of the labor crowd. I think the Corporation would be less vulnerable if they had their own system of collective bargaining. It is a case of a responsible benevolent autocracy versus an irresponsible malignant autocracy. As between the two I would choose ——— neither but instead a responsible democracy."

24. Diary entry, Oct. 1, 1919; A. A. ("Fred") Corey to Dickson, Oct. 20, 1919.

25. Dickson to Wm. G. Mather, Dec. 6, 1919, Midvale folder 5. For the change in his explanation, see address to labor conference, New York, Feb. 16, 1920. See also, address, Congregational Club of New York, Feb. 9, 1920, box 3, WBD Papers.

26. James J. Campbell (asst. sec'y., Carnegie Steel Co.) to Dickson, Dec. 16, 1918 [*sic*], Dickson's reply, dated Dec. 6, 1928 [*sic*], begins "Pardon the delay in acknowledging yours of December 16th." Since Dickson in the course of the letter refers to Mussolini in Italy, the date 1928 rather than 1918 or 1919 would seem to be right. Whether Campbell's letter was misdated or Dickson waited ten years to reply is not clear. Copies of the correspondence will be found in "Memoirs," Chapter 15.

27. Interchurch World Movement Committee interviews, box 26, Saposs Labor Collection. Interviews with Johnstown workers Frank Friedhoof, "American Worker," June Williams, and Mr. Hip.

28. Ibid., Mrs. Muffit, Frank Friedhoff, Mr. Hip, Mr. Stuart, (Fairfield Ave.) and Ed Friedhoff.

29. Ibid., Mr. Stuart (828 Wood Street), Mr. Hip, June Williams, and Mr. Stuart (Fairfield Ave.).

30. *Johnstown Tribune*, Jan. 13, 1920.

31. Ibid., Aug. 5, 1920.

32. Corey to Dickson and Chadbourne, Jan. 17, 1920, Midvale folder 5.

33. Ibid., Dickson to Corey, Jan. 19, 1920.

34. Ibid., Dickson to Corey, Jan. 26 and Apr. 26, 1920.

35. Ibid., Corey to Dickson, Apr. 28, 1920. For Corey's salary negotiations, see diary entries, Mar. 22–24, 1919. For workers' incomes, see Midvale annual report, *Iron Age*, 109 (Mar. 10, 1921): 676.

36. Midvale annual report, *Iron Age*, 111 (Mar. 9, 1922): 657; diary entry, July 15, 1921.

37. Diary entry, Aug. 17, 1922; Corey to Dinkey, Aug. 11, 1922, enclosing photo, Midvale folder 5.

38. "R. B." to Dickson, Jan. 2, 1920, Midvale folder 5.

39. For example, see "Some Twentieth Century Problems," *AAP&SS*, 85 (Sep. 1919); undated notes and Dickson to Dinkey, July 20, 1920, Midvale folder 4, box 7, WBD Papers.

40. Diary entry, Feb. 3, 1922.

41. *Iron Age*, 106 (Nov. 11, 1920): 1253–254.

42. Diary entries, Oct. 20, 1921; Aug. 9, 1922 and July 20, 1923 (quoted).

43. "A Plain American" [Dickson] to Editor, *New York Times*, July 25, 1922, Letters to Editors folder, box 5, WBD Papers; diary entries, July 27, Aug. 3 and 4, 1922.

44. Copy of address, "Memoirs," Chapter 15. This particular address to the ERP representatives illustrates how Dickson's reading influenced his actions. See diary entry, June 17, 1919 (two months prior to the ERP meeting): "Finished Thoreau's 'Walden.' I agree with his philosophy in the main and intend to follow it as far as practicable i.e. in the elimination of non-essentials and in getting away from the 'tyranny of things' and of petty conventions."

45. Diary entry, Feb. 1, 1921; address, Feb. 28, 1922, box 3, WBD Papers; diary entries, Aug. 3, 1922 and Feb. 19, 26, and 27, 1923.

46. F. M. Mansfield, Sr., Audit Clerk, Cambria Steel Co. (and a member of the plant conference committee), "Midvale Plan of Representation," received by Dickson, Apr. 16, 1920, Midvale folder 5.

47. Based on quarterly reports of Midvale Steel, reported in financial section, *New York Times*, passim, 1916–1923.

48. As noted in Dickson's diary, Jan. 5, 1932.

49. Memo enclosed in Dickson to A. A. ("Fred") Corey, Dec. 1, 1920, Midvale folder 5.

50. *Iron Age*, 107 (Feb. 10, 1921): 400.

51. Memo enclosed in Dickson to Corey, Dec. 1, 1920, Midvale folder 5.

52. Diary entries, Jan. 21 and 28, 1921.

53. Memo enclosed in Dickson to Corey, Dec. 1, 1920, Midvale folder 5.

54. Diary entry, Jan. 6, 1921.

55. Diary entries, Feb. 1 and 4, 1921. Dickson's remarks are attached to the diary entry.

56. *Iron Age*, 107 (Feb. 17, 1921): 459.

57. Ibid., pp. 468, 527, 654, 726, 804, 866, 993, 998, 1196; *New York Times*, Apr. 25 and June 26, 1921.

58. *New York Times*, Apr. 7, 1921; diary entries, Mar. 17, Apr. 15, and 19, 1921 and passim, May 1922.

59. Diary entries, Nov. 13, 1921; Feb. 9 and 25, 1922.

60. Diary entries, Nov. 3, 15, and 21, 1921; Jan. 20, 21 and Mar. 10, 1922.

61. Diary entries, Nov. 23, 1921 and Mar. 7 and 8, 1922.

62. Diary entries, Mar. 8 and 16, 1922.

63. Diary entries, July 7, Sep. 28, and Oct. 2, 1922.

64. Family Scrapbook 2.

65. Diary entries, Apr. 8 and 10, 1919; Apr. 5, 1922 and passim, 1921–1922. *Iron Age*, 108 (Dec. 8, 1921): 1492–493, discusses the seven-company merger.

66. Family Scrapbook 2.

67. Diary entries, passim, March-July, 1922. See especially July 25, 1922.

68. Diary entries, Aug. 29 and Sep. 1, 1922.

69. Diary entries, Sep. 27 and 29, Nov. 2 and 24, 1922.

70. Dickson to Paul U. Kellogg, July 11, 1941, Kellogg folder, box 2, WBD Papers.

71. Diary entries, Mar. 23, 29, and 30, 1923.

72. Diary entry, Mar. 31, 1923.

73. Diary entry, Oct. 18, 1922.

74. In Oct. 1922, Dickson made plans to rent his Montclair residence. It was at that time that Mamie was upset and that Dickson expressed relief at giving up the place. See diary entries, Oct. 20, 1922 and Apr. 6 and 7, 1923.

75. Diary entries, May 24, 1923 and Dec. 23, 1924.

76. Diary entries, Dec. 31, 1924; Sep. 20, 1923; June 5, 1924 and Nov. 6, 1926.

Chapter 7. What Might Have Been . . .

1. Dickson, "Eight-Hour Day and Six-Day Week in the Continuous Industries," *American Labor Legislation Review*, 7 (Mar. 1917). The address was widely quoted and commented on. See, for example, *Iron Trade Review*, Jan. 11, 1917, pp. 124 and 143–46.

2. Diary entries, Sep. 24, 1918 and July 30, 1919, box 1, WBD Papers.

3. Edwin S. Mills to Dickson (enclosing clipping), Dec. 10; Dickson to Mills, Dec. 13, 1920, Midvale folder 7, box 7, WBD Papers.

4. Ibid., Baker to Dickson, Dec. 13; Dickson to Baker, Dec. 14; Baker to Dickson, Dec. 20, 1920.

5. Diary entries, Jan. 27 and Mar. 8, 1921.

6. For Gary's views and Dickson's comments, see "Memoirs," Chapter 10, box 2, WBD Papers.

7. Interchurch World Movement, *Report on the Steel Strike of 1919* (New York: Harcourt, Brace and Howe, 1920), chapter 1–3.

8. Kenneth E. Trombley, *The Life and Times of a Happy Liberal, A Biography of Morris Llewellyn Cooke* (New York: Harper, 1954), pp. 97–98; Kellogg to Dickson, May 6, 1936, U.S. Steel folder 6, box 7, WBD Papers.

9. *Survey*, 45 (Mar. 5, 1921): 783–84.

10. Horace B. Drury, "The Three-Shift System in the Steel Industry," *Bulletin of the Taylor Society*, 6 (Feb. 1921): 2–49. Despite earlier experiments with the three-shift system, the eight-hour shift in steel came to Western Europe in the wake of World War I. Germany adopted it in October 1918 during the revolution that overturned the Kaiser's government. Belgium, France, and Britain followed in late 1918 and 1919. See Gerald D. Feldman, *Iron and Steel in the German Inflation, 1916–1923* (Princeton: Princeton University Press, 1977), pp. 338–39; Duncan L. Burn, *The Economic History of Steelmaking, 1867–1939* (Cambridge: Cambridge University Press, 1940), pp. 354–55; David Murray, *Steel Curtain: A Biography of the British Iron and Steel Industry* (London: Pall Mall Press, 1959), p. 85.

11. *Survey*, 45: 783.

12. *Iron Age*, 107: 635–38.

13. Ibid., 107 (Mar. 10, 1921): 646; *New York World*, Mar. 8, 1921. *Iron Age*, 107 (Apr. 21, 1921): 1043 f.; Brody, *Steelworkers*, pp. 272–73.

14. Except where otherwise indicated, the account of the Harding administration's intervention in the hours controversy is based on Herbert Hoover, *The Memoirs of Herbert Hoover, 1920–1933, The Cabinet and the Presidency*, (New York; Macmillan, 1952), pp. 103–05; Robert K. Murray, *The Harding Era, Warren G. Harding and His Administration* (Minneapolis: University of Minnesota Press, 1969), pp. 234–38; and Robert H. Zieger, *Republicans and Labor, 1919–1929* (Lexington: University of Kentucky Press, 1969), pp. 98–108. Zieger's excellent account is the fullest.

15. Kellogg to Dickson, May 6, 1936, U.S. Steel folder 6.

16. Diary entry, Mar. 14, 1922.

17. A copy of Gary's remarks can be found in "Memoirs," Chapter 10.

18. Lindsay to Hoover, May 11, 1923, copy in Gary folder, box 6, WBD Papers.

19. Ibid.; Dickson to Samuel Conover, et al, Apr. 12, 1935, "Memoirs," Chapter 2.

20. Dickson to E. J. Buffington, Apr. 9, 1936, U.S. Steel folder 6.

21. Diary entry, May 19, 1922.

22. Murray, *Harding Era,* pp. 236–37, quotes Harding's letter to the Gary committee.

23. Zieger, *Republicans and Labor,* pp. 103–04; Kellogg to Dickson, May 6, 1936, U.S. Steel folder 6.

24. Feldman, *Iron & Steel in German Inflation,* pp. 339–40 and 440–42.

25. *Baltimore Sun,* May 27, 1923; Report of the American Iron & Steel Institute, May 25, 1923, typewritten copy in WBD Papers.

26. Brody, *Steelworkers,* pp. 273–74; *New York American,* June 6, 1923.

27. Drury to Cooke, undated, copy in Gary folder, box 6.

28. Murray, *Harding Era,* p. 273. A copy of Harding's letter can be found in "Memoirs," Chapter 10.

29. Zieger, *Republicans and Labor,* p. 105, poses the dilemma of the steelmen. A copy of the Iron & Steel Institute's reply to Harding can be found in "Memoirs," Chapter 10.

30. Murray, *Harding Era,* pp. 237–38; Zieger, *Republicans and Labor,* p. 106.

31. Gary interview, *New York World,* Aug. 12, 1923.

32. Original Poetry folder, box 3, WBD Papers.

33. Brody, *Steelworkers,* p. 274 n.; "Havoc Wrought by the Shorter Work Day in Steel," *Survey,* 57 (Jan. 1, 1927): 465. See also, "Chairman Gary as Prophet," *New York World,* Mar. 25, 1924.

34. Diary entry, Oct. 25, 1923.

35. Dickson to Samuel Conover, et al. Apr. 12, 1935; Dickson to Taylor, Aug. 24, 1920 "Memoirs," Chapter 10.

36. Dickson to Arthur H. Young, Mar. 23, 1936. See also, Dickson memo, Apr. 11, 1937, "Industrial Democracy Scorned, Autocracy of Capital Defeated, Autocracy of Labor Triumphant," containing copies of several letters written to U.S. Steel officials in 1935 and 1936, "Memoirs," Chapter 10.

37. Dickson to President Franklin D. Roosevelt (telegram), June 14, 1934. Dickson subsequently dined with the president at the White House on Dec. 7, 1934, Executive Correspondence folder, box 5, WBD Papers.

38. Concluding remarks, "Memoirs," Chapter 10.

39. Dickson to Congressman E. K. Hall, June 11, 1932, Congressional folder, box 5, WBD Papers.

40. Ibid., Dickson to Senator Pat Harrison, Feb. 3, 1933.

41. Copies of the pamphlet (or portions thereof) variously entitled *Democracy in Industry, A Social Crisis,* and *Democracy at the Crossroads,* can be found in box 3, WBD Papers.

42. Morgan to Dickson, June 3, 1935; Buffington to Dickson, Mar. 13, 1936. See also, Dickson to John W. Davis, May 26, 1938, U.S. Steel folder 6.

43. Diary entries, Dec. 9, 1918; Jan. 16, 1922 and Mar. 7, 1923.

44. Diary entry, Apr. 18, 1922.

45. Diary entry, Nov. 26, 1919.

46. Diary entry, Aug. 11, 1919.

47. Diary entries, Oct. 27, 1922 and Aug. 16, 1927.

48. Diary entry, Aug. 15, 1931.

49. See Ellis W. Hawley, "Herbert Hoover, the Commerce Secretariat, and the Vision of an 'Associative State'," *Journal of American History*, 61 (June 1974): 140.

50. Livesay, *Carnegie and Big Business*, pp. 165–69.

51. Hessen, *Steel Titan*, pp. 115–16.

52. For a description of modern techno-structures, see John K. Galbraith, *The New Industrial State* (Boston: Houghton Mifflin, 1967), pp. 74–84.

53. Garraty, "U.S. Steel vs. Labor," p. 31, claims that ending unnecessary Sunday labor was Gary's first assertion of authority after being designated chief executive officer of U.S. Steel. See also, Brody, *Steelworkers*, pp. 155–56.

54. Stanley Committee, *Hearings*, 1: 79.

55. Gulick, *Labor Policy of U.S. Steel*, pp. 185 and 189. On his copy of the book, Dickson wrote, "This book should be preserved. It proves conclusively that all the criticism I have made of Chairman Elbert H. Gary and the partners of J. P. Morgan & Co. were more than justified." Box 7, WBD Papers.

56. Carroll E. French, a contemporary student of ERPs, writing in 1923, offers insights into company unionism from the perspective of the early 1920s. He defined company unionism as a system of "collective dealing" rather than of collective bargaining and rejected it as a form of "Industrial Democracy." At the same time, he saw company unions as a possible compromise between regular unions, which companies would not accept, and the open shop. *The Shop Committee in the U.S.*

57. Quoted, David Brody, "The Rise and Decline of Welfare Capitalism," *Change and Continuity in Twentieth Century America: The 1920's*, ed. John Braeman, Robert H. Bremner, and David Brody. (Columbus: Ohio State University Press, 1968), p. 162.

58. Ibid.

List of Sources

Primary

Manuscript

Library of Congress, Washington, D.C. Andrew Carnegie Papers.

National War Labor Board. "Employees vs. Midvale Steel and Ordnance Co." Case file #129. National Archives, Record Group 2, Federal Records Center, Suitland, Maryland.

The Pennsylvania State University Library, University Park, Pennsylvania. William Brown Dickson Papers.

Notes for "Steel and Democracy, Memoirs of William Brown Dickson."

Manuscript Diaries, 1907–1912, 1918–1928, 1932.

Family Scrapbooks.

Correspondence.

Clippings.

Copies of addresses, essays, pamphlets, poetry, etc.

Plan of Representation of the Employees of the Midvale Steel & Ordnance Company, Cambria Steel Company and Subsidiary Companies.

The Steelworkers. Midvale Steel & Ordnance Company Magazine, January 1919.

[John O'Brien]. *Detailed Survey of Plan of Representation of the Midvale Steel & Ordnance Company, New York City, Philadelphia, Nicetown and Johnstown.* N.p., [1919].

F. M. Mansfield, Sr. Unpublished manuscript, "Midvale Plan of Representation."

State Historical Society of Wisconsin, Madison, Wisconsin. "Organizing the Steel Workers" [by David A. Saposs]. Unpublished manuscript.

University Library, Pittsburgh. "History of the Strike in Johnstown." Heber Blankenhorn Papers, 1919–1937. Archives of Industrial Society [by George Soule].

Government Documents

U.S. Congress. Senate. *Report on Conditions of Employment in the Iron and Steel Industry.* By Neill, Charles P. (For the Bureau of Labor). 62d Cong., 1st Sess., Senate Document 110. 4 vols. Washington, D.C.: U.S. Government Printing Office, 1911–1913.

U.S. Dept. of Commerce, Bureau of the Census. *Historical Statistics of the United*

States, Colonial Times to 1970. 2 vols. Washington, D.C.: U.S. Government Printing Office, 1975.

U.S. Dept. of Labor, Bureau of Labor Statistics. *National War Labor Board.* Bulletin #287. Washington, D.C.: U.S. Government Printing Office, 1922.

U.S. Congress. House. Committee on the Investigation of the United States Steel Corporation. *Hearings.* 62d Cong., 2d Sess. 8 vols. Washington, D.C.: U.S. Government Printing Office, 1911–1913.

———. *Investigation of the Employment of Pinkerton Detectives in Connection with the Labor Troubles at Homestead, Pa.* 52d Cong., 1st Sess., Miscellaneous Document 335. Washington, D.C.: U.S. Government Printing Office, 1897.

U.S. Congress. Senate, Committee on Education and Labor. *Hearings on Eight hours for Laborers in Government Contracts.* 57th Cong., 2d Sess., Senate Document 141. Washington, D.C.: U.S. Government Printing Office, 1903.

———. Committee on Education and Labor. *Report Upon the Relations Between Capital and Labor and Testimony Taken by the Committee.* 4 vols. Washington, D.C.: U.S. Government Printing Office, 1885.

Minutes, Pamphlets, and Reports

Address of the President and Papers Delivered at the American Iron & Steel Institute, First Formal Meeting, Waldorf-Astoria, New York City, Oct. 14, 1914. N.p., n.d.

The Annual Statistical Report of the American Iron & Steel Institute for 1914. Philadelphia, 1915.

Dickson, William B. *Paper Read to 28th Annual Reunion and Dinner of Carnegie Veteran Association, December 13, 1929.* N.p., n.d.

———. *The Problem of Organized Labor and Organized Capital.* Montclair, N.J.: N.p., 1915.

Fitch, John A. *Hours of Labor in the Steel Industry, A Communication to 15,000 Stockholders of the United States Steel Corporation.* N.p., [1912].

Metal Statistics 1978. New York: Fairchild Publications, 1978.

Statistics of the American and Foreign Iron Trade for 1912. Philadelphia: Bureau of Statistics, American Iron & Steel Institute, 1913.

United States Steel Corporation. *Action of United States Steel Corporation Upon Recommendations of Stockholders' Committee.* N.p., n.d.

United States Steel Corporation. *Copies of Letters Received From Stockholders in Answer to C. M. Cabot Circular Letter of March 26, 1912.* N.p., n.d.

———. *Minutes of Annual Meeting of Stockholders.* N.p., 1910–1923.

———. *Report Of Committee of Stockholders.* New York, April 15, 1912.

Other Primary Sources

Beyer, David S. "Safety Provisions in the United States Steel Corporation." *Survey,* 24 (May 7, 1910): 205–36.

Bolling, Raynal C. "Rendering Labor Safe in Mine and Mill." *American Iron & Steel Institute Yearbook, 1912,* pp. 106–13.

———. "Results of Voluntary Relief Plan of United States Steel Corporation." *Annals of the American Academy of Political and Social Sciences,* 38 (July–December 1911): 35–44.

Bull, R.A. "The Twelve-Hour Shift in the Steel Foundry." *Iron Age,* 90 (Oct. 3, 1912): 808–09.

Calder, John. "Five Years of Employee Representation Under 'The Bethlehem Plan.' " *Iron Age*, 111 (June 14, 1923): 1689–696.
Carnegie, Andrew. *The Gospel of Wealth and Other Timely Essays*. Edited by Edward C. Kirkland. Cambridge, Mass.: Belknap Press, 1962.
"Cooperation and Conciliation in the Steel Industry." *Iron Age*, 80 (Nov. 28, 1907): 1549.
"Democracy in Steel-Making." *Iron Age*, 102 (Sep. 26, 1918): 764.
Dickson, William B. "Can American Steel Plants Afford an Eight-Hour Turn?" *Survey*, 32 (Jan. 3, 1914): 376.
______. "Eight-Hour Day and Six-Day Week in the Continuous Industries." *American Labor Legislation Review*, 7 (March 1917): 3–15.
______. "Getting Our Men to Help Us Manage." *System*, 35 (June 1919): 1041–44.
______. "The Interrupted Dream." *Survey*, 26 (June 3, 1911): 344.
______. "The Kaiser's Vision." *Manufacturers Record*, Feb. 21, 1918.
______. "New Jersey Employers' Liability and Workmen's Compensation Law." *Annals of the American Academy of Political and Social Sciences*, 38 (July-December 1911): 218–24.
______. "Some Twentieth Century Problems." *Annals of the American Academy of Political and Social Sciences*, 85 (Sep. 1919): 12–27.
______. "Some War Problems of the Steel Trade." *Iron Age*, 102 (Oct. 24, 1918): 1015–016.
______. War Tax and Bond Issues." *Iron Age*, 101 (June 20, 1918): 1610–611.
Dickson, William B., comp. *History of the Carnegie Veteran Association*. Montclair, N.J.: Mountain Press, 1938.
Drury, Horace B. "The Three-Shift System in the Steel Industry." *Bulletin of the Taylor Society*, 6 (Feb. 1921): 2–49.
Edge, Walter E. "New Jersey Employers' Liability Act." *Annals of the American Academy of Political and Social Sciences*, 38 (July-December 1911): 225–29.
Fitch, John A. "Holding Fast to the Twelve Hour Day." *Survey*, 30 (May 3, 1913): 165–66.
______. "Old Age at Forty." *American Magazine*, 71 (March 1911): 655–64.
______. *The Steel Workers*. New York: Charities Publication Committee, 1911.
Gould, E. R. L. *The Social Condition of Labor*. Johns Hopkins University Studies in Historical and Political Studies. Herbert B. Adams, editor. Baltimore: The Johns Hopkins Press. 1893.
Hoover, Herbert C. *The Memoirs of Herbert Hoover, 1920–1933, The Cabinet and The Presidency*. New York: Macmillan, 1952.
Interchurch World Movement. *Report on the Steel Strike of 1919*. New York: Harcourt, Brace and Howe, 1920.
"Iron and Steel Manufacturers Do Honor to Chairman Gary." *Iron Age*, 84 (Oct. 21, 1909): 1217–222.
Jones, William. "On the Manufacture of Bessemer Steel and Steel Rails in the United States." *Journal of the Iron & Steel Institute, 1881*, pp. 129–40.
"Judge Gary in England." *Iron Age*, 84 (Sep. 9, 1909): 786–87.
King, C.D. *Seventy-five Years of Progress in Iron and Steel*, New York: American Institute of Mining and Metallurgical Engineers, 1948.
"Midvale Pension and Home-Building Plans." *Iron Age*, 104 (July 17, 1919): 182–83.
"Midvale Plan of Employee Representation." *Iron Age*, 102 (Sep. 26, 1918): 763.
"Midvale Plan of Employee Representation." *Iron Age*, 102 (Oct. 3, 1918): 834–35.
"Midvale Plan of Representation Tested." *Iron Age*, 105 (Mar. 18, 1920): 815.
Olds, Marshall. *Analysis of the Interchurch World Movement Report on the Steel Strike*. New York: G. P. Putnam's Sons, 1923.

Porter, John Jermain. "The Manufacture of Pig Iron." In *The ABC of Iron and Steel*. 2d ed. Edited by A. O. Backert. Cleveland: Penton Publishing Co., 1917.

Schwedtman, Ferdinand C. "Principles of Sound Employers' Liability Legislation." *Annals of the American Academy of Political and Social Sciences*, 38 (July-December 1911): 202–04.

Schwedtman, Ferdinand C., and Emery, James A. *Accident Prevention and Relief*. New York: National Association of Manufacturers, 1911.

"Stable Steel Pries Defended by Judge Gary." *Iron Age*, 82 (Dec. 17, 1908): 1794–795.

"Steel Prices to be Maintained." *Iron Age*, 81 (May 28, 1909): 1710–711.

"The Tribute to Judge Gary." *Iron Age*, 84 (Oct. 21, 1909): 1246.

Wilson, Edward. "The Organization of an Open Shop Under the Midvale Plan." *Annals of the American Academy of Political and Social Sciences*, 85 (Sep. 1919): 214–19.

Secondary Articles and Books

Allen, Frederick Lewis. *The Great Pierpont Morgan*. New York: Harper and Brothers, 1949.

Asher, Robert. "Painful Memories: The Historical Consciousness of Steel Workers and the Steel Strike of 1919." *Pennsylvania History*, 45 (January 1978): 61–86.

Bemis, Edward W. "The Homestead Strike." *Journal of Political Economy*, 2 (June 1894): 369–96.

Brandes, Stuart D. *American Welfare Capitalism, 1880–1940*. Chicago: University of Chicago Press, 1976.

Bridge, James H. *The Inside History of the Carnegie Steel Company*. New York: Aldine Book Co., 1903.

Brody, David. *Labor in Crisis: The Steel Strike of 1919*. Philadelphia, J. B. Lippincott, 1965.

———. "The Rise and Decline of Welfare Capitalism." *Change and Continuity in Twentieth Century America: The 1920's*. Edited by John Braeman, Robert H. Bremner, and David Brody. Columbus: Ohio State University Press, 1968.

———. *Steelworkers in America, the Nonunion Era*. Cambridge, Mass.: Harvard University Press, 1960.

Burn, Duncan L. *The Economic History of Steelmaking, 1867–1939*. Cambridge: Cambridge University Press, 1940.

Chambers, Clarke A. *Paul U. Kellogg and the Survey: Voices for Social Welfare and Social Justice*. Minneapolis: University of Minnesota Press, 1971.

"Charles M. Cabot—Industrial Reformer." *Survey*, 35 (Oct. 2, 1915): 27–29.

Cochran, Thomas. "Gary." *Encyclopedia of American Biography*. Edited by John A. Garraty. New York: Harper and Row, 1974.

Copley, Frank Barkley. "A Great Corporation Investigates Itself." *American Magazine*, 74 (Oct. 1912): 643–54.

Cotter, Arundel. *United States Steel: A Corporation With a Soul*. Garden City, N.Y.: Doubleday, Page & Company, 1921.

Cuff, Robert D. *The War Industries Board: Business-Government Relations During World War I*. Baltimore: The Johns Hopkins University Press, 1973.

Dunn, Robert W. *Company Unions*. New York: Vanguard Press, 1927.

Faulkner, Harold U. "Gary, Elbert Henry." *Dictionary of American Biography*, 7: 175–176.

Feldman, Gerald D. *Iron and Steel in the German Inflation, 1916–1923*. Princeton: Princeton University Press, 1977.

French, Carroll E. *The Shop Committee in the United States*. Johns Hopkins University

Studies in History and Political Science, vol. 41, #2. Baltimore: The Johns Hopkins Press, 1923.
Galambos, Louis. "The Emerging Organizational Synthesis of Modern American History." *Business History Review*, 44 (Autumn 1970): 279–90.
Galbraith, John Kenneth. *The Age of Uncertainty*. Boston: Houghton Mifflin, 1977.
______. *The New Industrial State*. Boston: Houghton Mifflin, 1967.
Garraty, John A. *Right-Hand Man, The Life of George W. Perkins*. New York: Harper and Brothers, 1957.
______. "The United States Steel Corporation Versus Labor: The Early Years." *Labor History*, 1 (Winter 1960): 3–38.
Gulick, Charles A. *Labor Policy of the United States Steel Corporation*. New York: Columbia University Press, 1924.
Hard, William. "Making Steel and Killing Men." *Everybody's Magazine*, 17 (November 1917): 579–91.
Harvey, George. *Henry Clay Frick: The Man*. New York: Charles Scribner's Sons, 1928.
"Havoc Wrought by the Shorter Work Day in Steel." *Survey*, 57 (Jan. 1, 1927): 465.
Hawley, Ellis W. "The Discovery and Study of a 'Coporate Liberalism'." *Business History Review*, 52 (Autumn 1978): 309–20.
______. "Herbert Hoover, the Commerce Secretariat, and the Vision of an 'Associative State'." *Journal of American History*, 61 (June 1974): 116–40.
Heald, Morrell. *The Social Responsibilities of Business, Company and Community, 1900–1960*. Cleveland: Case Western Reserve University Press, 1970.
Hendrick, Burton J. *The Life of Andrew Carnegie*. 2 vols. Garden City, N.Y.: Doubleday, Doran and Co., 1932.
Hessen, Robert. "The Bethlehem Steel Strike of 1910." *Labor History*, 15 (Winter 1974): 3–18.
______. *Steel Titan: The Life of Charles M. Schwab*. New York: Oxford University Press, 1975.
Hill, Charles. "Fighting the Twelve-Hour Day in the American Steel Industry." *Labor History*, 15 (Winter 1974): 19–35.
Kipnis, David. *The Powerholders*. Chicago: University of Chicago Press, 1976.
Kolko, Gabriel. *The Triumph of Conservativism*. New York: Free Press of Glencoe, 1963.
Livesay, Harold C. *Andrew Carnegie and the Rise of Big Business*. Boston: Little Brown, 1975.
Lubove, Roy. *The Struggle for Social Security, 1900–1935*. Cambridge, Mass.: Harvard University Press, 1968.
McClymer, John T. "The Pittsburgh Survey, 1907-1914: Forging an Ideology in the Steel District." *Pennsylvania History*, 41 (April 1974): 169–86.
Miller, Frank Woodward. "Swissvale Then and Now, A Brief Historical Sketch." *The Twentieth Anniversary of Swissvale Borough*. Swissvale: N.p., 1918.
Murray, David. *Steel Curtain: A Biography of the British Iron and Steel Industry*. London: Pall Mall Press, 1959.
Murray, Robert K. *The Harding Era, Warren G. Harding and His Administration*. Minneapolis: University of Minnesota Press, 1969.
National Industrial Conference Board. Research Report #21. *Works Councils in the United States. . . .* Boston, 1919.
Nelson, Daniel. *Managers and Workers: Origins of the New Factory System in the United States 1880–1920*. Madison: University of Wisconsin Press, 1975.
Page, Kirby. "Gold from Steel." *The World Tomorrow*, April 12, 1933.

Paine, Ralph D. "William Ellis Corey." *World's Work*, 6 (October 1903): 4025–027.

Satterlee, Herbert L. *J. Pierpont Morgan, An Intimate Portrait.* New York: Macmillan, 1939.

Selekman, Ben M., and Van Kleeck, Mary. *Employees' Representation in Coal Mines, A Study of the Industrial Representation Plan of the Colorado Fuel & Iron Co.* New York: Russell Sage Foundation, 1924.

Sklar, Martin J. "Woodrow Wilson and the Political Economy of Modern United States Liberalism." *Studies on the Left*, 1 (1960): 17–47.

Taft, Philip. *Organized Labor in American History.* New York: Harper and Row, 1964.

Tarbell, Ida. *The Life of Elbert H. Gary.* New York: D. Appleton, 1925.

Taylor, Albion Guilford. *Labor Policies of the N.A.M.* Urbana: The University of Illinois, 1928.

Trombley, Kenneth E. *The Life and Times of a Happy Liberal, A Biography of Morris Llewellyn Cooke.* New York: Harper, 1954.

Urofsky, Melvin I. *Big Steel and the Wilson Administration.* Columbus: Ohio State University Press, 1969.

Wall, Joseph Frazier. *Andrew Carnegie.* New York: Oxford University Press, 1970.

Weinstein, James. *The Corporate Ideal in the Liberal State: 1900–1918.* Boston: Beacon Press, 1968.

Wiebe, Robert H. *Businessmen and Reform: A Study of the Progressive Movement.* Cambridge, Mass.: Harvard University Press, 1962.

———. *The Search for Order, 1877–1920.* New York: Hill and Wang, 1967.

Wolff, Leon. *Lockout: The Story of the Homestead Strike, 1892.* New York: Harper and Row, 1965.

Zieger, Robert H. *Republicans and Labor, 1919–1929.* Lexington: University of Kentucky Press, 1969.

Index